ELECTRICAL AND OPTICAL WAVE-MOTION

THE MATHEMATICAL ANALYSIS OF ELECTRICAL AND OPTICAL WAVE-MOTION

ON THE BASIS OF MAXWELL'S EQUATIONS

BY

H. BATEMAN, M.A., Ph.D.

Late Fellow of Trinity College, Cambridge;
Johnston Research Scholar, Johns Hopkins University, Baltimore

Cambridge:
at the University Press
1915

CAMBRIDGE
UNIVERSITY PRESS

University Printing House, Cambridge CB2 8BS, United Kingdom

Cambridge University Press is part of the University of Cambridge.

It furthers the University's mission by disseminating knowledge in the pursuit of education, learning and research at the highest international levels of excellence.

www.cambridge.org
Information on this title: www.cambridge.org/9781316626122

© Cambridge University Press 1915

This publication is in copyright. Subject to statutory exception and to the provisions of relevant collective licensing agreements, no reproduction of any part may take place without the written permission of Cambridge University Press.

First published 1915
First paperback edition 2016

A catalogue record for this publication is available from the British Library

ISBN 978-1-316-62612-2 Paperback

Cambridge University Press has no responsibility for the persistence or accuracy of URLs for external or third-party internet websites referred to in this publication, and does not guarantee that any content on such websites is, or will remain, accurate or appropriate.

PREFACE

THIS book is intended as an introduction to some recent developments of Maxwell's electromagnetic theory which are directly connected with the solution of the partial differential equation of wave-motion. The higher developments of the theory which are based on the dynamical equations of motion are not considered at all. Even with this limitation the subject is a vast one, and to bring the work of perusing the literature within my power I have omitted an account of the modern theory of relativity which has been expounded very clearly in several recent publications.

For a thorough understanding of the present subject a very extensive knowledge of mathematics is necessary, but there are parts of the subject in which a reader with only a limited mathematical equipment may soon feel at home and perhaps do useful original work. With the idea of enabling such a reader to obtain a quick grasp of the nature of the subject and the results obtained, I have thought it advisable to state without proof a number of relations of which adequate demonstrations can only be obtained by means of complicated and difficult analysis. I have also endeavoured to keep the analysis as elementary as possible, but in some places where the work is perfectly straightforward a few details are omitted.

The book is far from being a complete treatise on the subject, for I have not given any existence theorems to show that the solutions of certain problems exist and are unique, and no attempt has been made to enter into the details of numerical computations. There are many parts of the subject indeed to which a pure mathematician might make useful additions; in particular, I might direct attention to p. 21, line 2, and p. 101, where there are one or two matters which require further discussion.

Chapter VIII and paragraph 5 contain some of my own contributions to the subject. At present there seem to be several different directions in which future developments may be made, and so it seems unwise to give a hasty judgment concerning the physical significance of the results. Ideas which naturally present themselves are that the aether can be regarded as built up from singular curves of the type considered in § 43, and that §§ 41 and 44 may throw some light on the question of the difference between positive and negative elementary electric charges. I hope to discuss an hypothesis relating to the first idea in a future note, but am unable to give any support at present to the second idea.

I gratefully acknowledge my indebtedness to Sir Joseph Larmor who read the manuscript before it was revised and made some helpful suggestions, to Prof. Ames who read the greater portion of the manuscript, to Prof. Morley and Mr Hassé who helped me with their advice and vigilance in reading the proof-sheets, and to the officers and staff of the University Press for their careful work and constant consideration shown in matters connected with the printing. For the correctness of the new formulae and examples I alone am responsible; if any errors are discovered I shall be grateful if my readers will inform me.

HARRY BATEMAN.

October, 1914.

ADDITIONS AND CORRECTIONS

p. 28. Formula (30) is due to Lamé. Cf. A. E. H. Love, *The Mathematical Theory of Elasticity*, 2nd edition, p. 55.

p. 101. An asymptotic expression for $T_m^n(s)$ when n is a large positive integer can be derived from a formula given by L. Fejér in 1909. This formula is accessible in a paper by O. Perron, *Arkiv der Mat. u. Phys.* (1914).

p. 118. The factor c in front of the double integrals should be omitted.

p. 120. Delete the minus sign in the second of equations (277).

p. 127. Line 8. This statement is incorrect, the equations are poristic, the special case is the only one which can occur.

p. 132. Line 20. On account of the porism just mentioned, the hope may be abandoned.

p. 150. Ex. 13. For equations (10) of § 5 read equations (2) of § 2.

p. 154. Ex. 24. The equation should read

$$\frac{\partial}{\partial a}\left[\frac{\sigma}{t-\tau}\cos(a-\epsilon)\frac{\partial V}{\partial a}+\sigma\sin(a-\epsilon)\frac{\partial V}{\partial t}\right]+\frac{\partial}{\partial\theta}\left[(t-\tau)\frac{\partial V}{\partial a}\right]$$
$$+\frac{\partial}{\partial t}\left[\sigma\sin(a-\epsilon)\frac{\partial V}{\partial a}+(t-\tau)\frac{\partial V}{\partial\theta}-\sigma(t-\tau)\cos(a-\epsilon)\frac{\partial V}{\partial t}\right]=0.$$

[illegible] AND CORRECTIONS.

p. [illegible] [illegible], R. H. Bolt, [illegible] Theory of [illegible], p. [illegible]

p. [illegible] [illegible] [illegible] positive [illegible] [illegible] [illegible] [illegible] [illegible] [illegible] generally [illegible] [illegible] [illegible]

p. 118. [illegible] [illegible] of the [illegible] should [illegible]

p. 130. [illegible] [illegible] instead of equation [illegible]

p. 137. [illegible] [illegible] the equation [illegible] [illegible] [illegible]

p. 138. [illegible] [illegible] of the [illegible] [illegible]

p. 138. [illegible] [illegible] equations (80) [illegible] equations [illegible]

p. 1[illegible]. [illegible] [illegible] [illegible]

[illegible]

CONTENTS

CHAPTER I

FUNDAMENTAL IDEAS

§ 1. The fundamental equations for free aether.

In Maxwell's electromagnetic theory the state of the aether in the vicinity of a point (x, y, z) at time t is specified by means of two vectors E and H which satisfy the circuital relations*

$$\operatorname{rot} H = \frac{1}{c}\frac{\partial E}{\partial t}, \quad \operatorname{rot} E = -\frac{1}{c}\frac{\partial H}{\partial t} \quad \text{..............(1)},$$

and the solenoidal or sourceless conditions

$$\operatorname{div} E = 0, \qquad \operatorname{div} H = 0.$$

If right-handed rectangular axes are used the symbol† rot H denotes the vector whose components are of type

$$\frac{\partial H_z}{\partial y} - \frac{\partial H_y}{\partial z},$$

the three components of H being H_x, H_y, H_z respectively. The symbol div H denotes the divergence of H, i.e. the quantity

$$\frac{\partial H_x}{\partial x} + \frac{\partial H_y}{\partial y} + \frac{\partial H_z}{\partial z}.$$

The vector E is called the *electric displacement* or *electric force* and H the *magnetic force*. The quantity c represents the

* The equations are written in the symmetrical form in which they were presented by O. Heaviside, *Electrical Papers*, Vol. 1, § 30, and H. Hertz, *Electric Waves*, p. 138. Sir Joseph Larmor points out that a set of equations equivalent to these was first used by MacCullagh in 1838 as a scheme consistently covering the whole ground of Physical Optics, *Collected Works* of James MacCullagh (1880), p. 145.

† We use here the units and notation employed in Lorentz's *The Theory of Electrons*, Ch. I, except that large letters are used to denote vectors and E is written in place of D. Many writers use the symbol *curl* instead of rot.

velocity of propagation of homogeneous plane waves and is commonly called the velocity of light; we shall assume it to be a constant, although in the most recent speculations it is treated as variable*.

Some of the modern writers on the theory of relativity maintain that the introduction of the idea of an aether is unnecessary and misleading. Their criticisms are directed chiefly against the popular conception of the aether as a kind of fluid or elastic solid which can be regarded as practically stationary while material and electrified particles move through it. This idea has been very helpful as it presents us with a vivid picture of the processes which may be supposed to take place, it also has the advantage that with its aid we can attach a meaning to the term absolute motion, but herein lies its weakness. Larmor, Lorentz and Einstein have shown, in fact, that the differential equations of the electron theory admit of a group of transformations which can be interpreted to mean that there is no such thing as absolute motion.

If this be admitted, the popular idea of the aether must be regarded as incorrect, and so if we wish to retain the idea of a continuous medium to explain action at a distance we must frankly acknowledge that the simplest description we can give of the properties of our medium is that embodied in the differential equations (1).

If we abandon the idea of a continuous medium in the usual sense only two ways of explaining action at a distance readily suggest themselves. We may either think of the aether as a collection of tubes or filaments attached to the particles of matter as in the form of Faraday's theory which has been developed by Sir Joseph Thomson and N. R. Campbell; or we may suppose that some particle or entity which belonged to an active body at time t belongs to the body acted upon at a later time $t + \tau$. From one point of view these two theories are the same, for if particles are continually emitted from an active

* A. Einstein, *Ann. d. Phys.* Vol. 35 (1911), p. 898; Vol. 38 (1912), pp. 355 and 443. M. Abraham, *Phys. Zeitschr.* (1912), pp. 1—5, 310—314, 793—797; *Ann. d. Phys.* (1912), pp. 444 and 1056; Fifth International Congress of Mathematicians, *Proceedings*, Vol. 2, p. 256.

body they will form a kind of thread attached to it. The first form of the theory is, however, more general than the second.

At present we are unable to form a satisfactory picture of the processes that give rise to, or are represented by, the vectors E and H. We believe, however, that some points may be made clear by studying the properties of solutions of our differential equations.

It will be seen from the investigations of Chapter VIII that the mathematical analysis connected with these equations is suitable for the discussion of three distinct theories of the universe, which may be described briefly as follows:—

Aether	*Matter*
Continuous medium.	Aggregates of discrete particles.
Discontinuous medium consisting of a collection of tubes or filaments.	An aggregate of discrete particles attached to the tubes.
Continuous medium.	An aggregate of discrete particles to which tubes are attached.

The last theory may be supposed to include that form of the emission theory of light in which small entities are projected from the particles of matter under certain circumstances and produce waves in the surrounding medium. This theory might be justly ascribed to Newton*.

For other theories of the aether the reader is referred to Prof. E. T. Whittaker's recent work† *A History of the Theories of the Aether*.

In the first part of this book the analysis is adapted almost entirely to the first theory, the high development of which we owe to the pioneer work of Maxwell, FitzGerald, Hertz, Rayleigh, Heaviside, J. J. Thomson, Lorentz and Larmor. The other theories have not yet received much attention but it is hoped

* A form of the theory in which the entities are electric doublets has been developed by W. H. Bragg and applied to the X and γ rays. *British Association Reports* (1911), p. 340.

† Dublin Univ. Press; Longmans, Green and Co. (1910).

that the analysis of Chapter VIII will lead to further developments so that a comparison can be made between the different theories. It is quite likely that one theory will be enriched by the developments of another.

§ 2. **Electromagnetic fields.**

For many purposes it is convenient to work with a complex vector* $M = H \pm iE$, where $i = \sqrt{-1}$ and the ambiguous sign $\pm$ is independent of the ambiguity which occurs in the determination of $\sqrt{-1}$. The differential equations (1) may then be replaced by the simpler equations

$$\operatorname{rot} M = \mp \frac{i}{c} \frac{\partial M}{\partial t}, \quad \operatorname{div} M = 0 \quad \text{.................(2).}$$

When a solution of these equations has been found a pair of vectors E and H satisfying equations (1) may be obtained by equating coefficients of the ambiguous sign. In working with an ambiguous sign it must be remembered that when two ambiguous signs are multiplied together the ambiguity is removed. The chief advantage in using the two independent ambiguities $\pm$ and $\sqrt{-1}$ is that we can assume that the vectors E and H are the real parts of expressions of the form $Ae^{i\omega t}$ and we are at liberty to equate the coefficients of either i or $\pm$ in any of our equations.

Definition. A solution of the differential equations (2) or (1), which provides us with single-valued vector functions E and H for each space-time point (x, y, z, t) belonging to a certain domain D, is said to define an electromagnetic field in the domain D.

Since the differential equations are linear the sum of any number of solutions is also a solution. The physical meaning of this is that when two electromagnetic fields are superposed, they are together equivalent to an electromagnetic field.

Two superposed electromagnetic fields can of course be related to one another in some way. When electromagnetic

* The use of a complex vector $H - iE$ is recommended by L. Silberstein, *Ann. d. Phys.* Vols. 22 and 24 (1907); *Phil. Mag.* (6), Vol. 23 (1912), p. 790. He does not, however, use the ambiguous sign.

waves fall upon an obstacle, a secondary disturbance is produced which depends in character upon the nature of both the primary waves and the obstacle.

We shall find that in some cases it is possible to find two fields in which the vectors (E, H), (E', H') are connected by the two relations embodied in the equation

$$(MM') \equiv M_x M_x' + M_y M_y' + M_z M_z' = 0 \quad \text{.........(3)}$$

for all values of (x, y, z, t) belonging to some domain.

When this is the case the fields are said to be *conjugate* within this domain.

If we use the notation

$$(M^2) = M_x^2 + M_y^2 + M_z^2,$$

we may write

$$(M^2) = (H^2) - (E^2) \pm 2i(EH) = I_1 \pm 2i\, I_2,$$

where I_1 and I_2 are two quantities which we shall call the invariants*. It is easy to see that when two conjugate fields are superposed the invariant I_1 for the total field is the sum of the invariants I_1 for the two component fields. Similarly for the invariant I_2.

When the invariants are zero over a given domain the field may be called self-conjugate for this region†.

§ 3. The flow of energy.

An entity whose volume density‡ ρ is a function of (x, y, z, t) will vary in a manner which can be described as a simple flow with component velocities (u, v, w) if the equation of continuity

$$\frac{\partial \rho}{\partial t} + \frac{\partial}{\partial x}(\rho u) + \frac{\partial}{\partial y}(\rho v) + \frac{\partial}{\partial z}(\rho w) = 0 \quad \text{...........(4)}$$

is satisfied. This equation implies in fact that there is no

* They are invariants for the group of linear transformations which leave the electromagnetic equations unaltered in form. Cf. H. Minkowski, *Gött. Nachr.* (1908); E. Cunningham, *Proc. London Math. Soc.* (2), Vol. 8 (1910), p. 89; H. Poincaré, *Rend. Palermo* (1906); M. Planck, *Ann. d. Phys.* Vol. 26 (1908). Other invariants are given by these authors.

† Silberstein calls it a pure electromagnetic wave.

‡ The limitations to which the idea of density is subject and the question of the continuity of the function ρ are discussed by J. G. Leathem, "Volume integrals and their use in physics," *Cambridge Mathematical Tracts* (1905).

creation or annihilation of the entity in the neighbourhood of (x, y, z, t).

Now it is easy to see that the equation of continuity is satisfied in virtue of equations (1) if we put

$$\rho = \tfrac{1}{2}(E^2) + \tfrac{1}{2}(H^2), \quad \rho u = c\,(E_y H_z - E_z H_y),$$

and two similar equations. We shall regard ρ in this case as the volume density of the *energy* contained in the electromagnetic field. The vector Σ whose components are of the type $c\,(E_y H_z - E_z H_y)$ can then be supposed to indicate the rate at which energy flows through the field. Since

$$\rho^2 (c^2 - u^2 - v^2 - w^2) = \tfrac{1}{4} c^2 (E^2 - H^2)^2 + c^2 (EH)^2,$$

it appears that energy travels through the field with a velocity which is less than the velocity of light. The velocity c is attained only in the case of a self-conjugate field.

The vector Σ was introduced by Prof. Poynting* and is usually called Poynting's vector. The idea of describing the transfer of energy in this way also occurred to Prof. Lamb before the publication of Poynting's work.

Example. Prove that the equation of continuity may be satisfied by putting

$$\rho u = \frac{1}{c}\frac{\partial\theta}{\partial t} E_x - \frac{\partial\theta}{\partial y} H_z + \frac{\partial\theta}{\partial z} H_y, \qquad \rho v = \frac{1}{c}\frac{\partial\theta}{\partial t} E_y - \frac{\partial\theta}{\partial z} H_x + \frac{\partial\theta}{\partial x} H_z,$$

$$\rho w = \frac{1}{c}\frac{\partial\theta}{\partial t} E_z - \frac{\partial\theta}{\partial x} H_y + \frac{\partial\theta}{\partial y} H_x, \qquad \rho c = -\frac{\partial\theta}{\partial x} E_x - \frac{\partial\theta}{\partial y} E_y - \frac{\partial\theta}{\partial z} E_z,$$

where θ is an arbitrary function. Obtain a similar solution by replacing E by H and H by $-E$.

§ 4. First solution of the fundamental equations.

Let us use the symbol Ωu to denote the Dalembertian† of u, viz.

$$\Omega u \equiv \Delta u - \frac{1}{c^2}\frac{\partial^2 u}{\partial t^2} = \frac{\partial^2 u}{\partial x^2} + \frac{\partial^2 u}{\partial y^2} + \frac{\partial^2 u}{\partial z^2} - \frac{1}{c^2}\frac{\partial^2 u}{\partial t^2},$$

* *Phil. Trans.* A, Vol. 175 (1884), p. 343. See also H. A. Lorentz, *The Theory of Electrons*, p. 22.

† This is the name suggested by Lorentz, *loc. cit.* p. 17. Many writers use Cauchy's symbol $\Box$ to denote the Dalembertian, but I think Ω is preferable because its form suggests a wave. Murphy's symbol Δ is also used here in place of the usual symbol ∇^2. E. B. Wilson and G. N. Lewis use the symbol $\diamondsuit^2 u$ to denote the Dalembertian of u. Cf. *Proc. Amer. Acad. of Arts and Sciences*, Vol. 48 (1912), p. 389.

and the symbol grad U to denote the vector whose components are

$$\frac{\partial U}{\partial x}, \frac{\partial U}{\partial y}, \frac{\partial U}{\partial z}$$

respectively. Let us also use $\Omega\Lambda$, where Λ is a vector with components $\Lambda_x, \Lambda_y, \Lambda_z$, to denote the vector whose components are $\Omega\Lambda_x, \Omega\Lambda_y, \Omega\Lambda_z$. The equation $\Omega u = 0$ will be called the *wave-equation* and a solution of this equation a *wave-function*. A vector function Λ will be said to satisfy the wave-equation when each of its components is a wave-function, i.e. if $\Omega\Lambda = 0$. We may now satisfy equations (1) and (2) by writing

$$M = \pm i \operatorname{rot} L \equiv \frac{1}{c}\frac{\partial L}{\partial t} + \operatorname{grad} \Lambda \quad \text{.............(5)},$$

where the *scalar potential* $\Lambda = \Psi \mp i\Phi$ and the *vector potential* $L = B \mp iA$ satisfy the equations

$$\Omega\Lambda = 0, \quad \Omega L = 0, \quad \operatorname{div} L + \frac{1}{c}\frac{\partial \Lambda}{\partial t} = 0 \quad \text{........(6)}.$$

The last three equations may be solved in a general way by writing

$$\left.\begin{aligned} L &= \frac{1}{c}\frac{\partial G}{\partial t} \pm i \operatorname{rot} G + \operatorname{grad} K \\ \Lambda &= -\operatorname{div} G - \frac{1}{c}\frac{\partial K}{\partial t} \end{aligned}\right\} \quad \text{.............(7)},$$

where the vector $G \equiv \Gamma \mp i\Pi$ and the scalar K satisfy the *wave-equation*

$$\Omega u = 0 \quad \text{........................(8)}.$$

The solution of equations (1) which is embodied in (5), (6) and (7) is a simple extension of Hertz's solution* and is suggested by Whittaker's solution† in terms of two scalar potentials. It is clear that the function K drops out when we differentiate to find M and so the electric and magnetic forces depend only on the vector G. The form of this vector indicates that the electromagnetic field can be regarded as the sum of two partial fields; one of these is derived from the vector Π and

* *Ann. d. Phys.* Vol. 36 (1888), p. 1. The general solution is given by Righi, *Bologna Mem.* (5), t. 9 (1901), p. 1; *Il Nuovo Cimento* (5), t. 2 (1901), p. 2. He finds suitable expressions for the vectors Π and Γ in a number of cases.

† *Proc. London Math. Soc.* Ser. 2, Vol. 1 (1903).

will be called a field of *electric type*, the other is derived from the function Γ and will be called a field of *magnetic type*.

This resolution of an electromagnetic field into two partial fields is analogous to the one used by H. M. Macdonald* in the study of the effect of an obstacle on a train of electric waves. The component fields are then of such a type that in one case the magnetic force normal to the obstacle vanishes over the surface of the latter, in the other case it is the electric force normal to the obstacle that vanishes. The same idea has been used recently by Mie† and Debye‡ in the treatment of the case of a spherical obstacle.

In Hertz's solution we have $\Gamma = 0$, $K = 0$ and Π has components $(0, 0, S)$.

The components of E and H are consequently given by the formulae

$$\left.\begin{aligned} E_x &= \frac{\partial^2 S}{\partial x \partial z}, & H_x &= \frac{1}{c}\frac{\partial^2 S}{\partial y \partial t} \\ E_y &= \frac{\partial^2 S}{\partial y \partial z}, & H_y &= -\frac{1}{c}\frac{\partial^2 S}{\partial x \partial t} \\ E_z &= \frac{\partial^2 S}{\partial z^2} - \frac{1}{c^2}\frac{\partial^2 S}{\partial t^2}, & H_z &= 0 \end{aligned}\right\} \quad \ldots\ldots\ldots(9).$$

Hertz uses Euler's wave-function§

$$S = \frac{1}{r}\sin \kappa (r - ct), \quad r^2 = x^2 + y^2 + z^2,$$

and obtains in this way a theory of his oscillator‖. The electric and magnetic forces become infinite at the origin which is therefore a singularity of the electromagnetic field. A singularity of this type is called a *vibrating electric doublet* and is regarded as the simplest model of a source of light or electromagnetic waves.

* *Electric Waves*, Ch. VI.

† *Ann. d. Phys.* Vol. 25 (1908), p. 382. ‡ *Ibid.* Vol. 30 (1909), p. 57.

§ Periodic solutions representing a disturbance sent out from n-fold poles had been used previously by H. A. Rowland and applied to the elucidation of optical phenomena. *Amer. Journal of Mathematics*, Vol. 6, p. 359; *Phil. Mag.* Vol. 17 (1884), p. 423. Cf. also Stokes, *Cambr. Phil. Trans.* (1849).

‖ To deal with the case in which the vibrations are damped we assume $S = \frac{1}{r} e^{+\nu(r-ct)} \sin \kappa (r - ct)$. Cf. K. Pearson and A. Lee, *Phil. Trans.* A, Vol. 193 (1900), p. 159.

The solutions of equations (1) which are obtained by superposing elementary solutions of this type are of great importance in physical optics.

When r is very great the most important terms in the expressions (9) are

$$E_x = -\frac{\kappa^2 xzs}{r^3}, \qquad H_x = \frac{\kappa^2 ys}{r^2},$$

$$E_y = -\frac{\kappa^2 yzs}{r^3}, \qquad H_y = -\frac{\kappa^2 xs}{r^2},$$

$$E_z = \frac{\kappa^2 (x^2 + y^2) s}{r^3}, \quad H_z = 0,$$

where $s = \sin \kappa (r - ct)$. All the other terms are of order $1/r^2$ or $1/r^3$. These expressions give

$$(EH) = 0, \qquad (E^2) - (H^2) = 0.$$

Hence at a very great distance from the origin the field is practically a self-conjugate field and so the energy travels with a velocity very nearly equal to the velocity of light. The expressions indicate that Poynting's vector is ultimately along the radius from the origin; now the electric and magnetic forces are at right angles to Poynting's vector and so the vibrations of the light-vector, whether we take it to be the electric or magnetic force, are at right angles to the radius. The waves sent out from the source have, then, the character of monochromatic light at a great distance from the origin*. The amplitudes of the vibrations at points on the same radius are proportional to the quantities $1/r$ when r is large, and so if the intensity of the light be measured by the square of the amplitude the inverse square law is fulfilled.

Since the electric force is ultimately at right angles to the radius there is no total charge associated with the singularity, for the charge is equal to the surface-integral of the normal electric force over a large sphere concentric with the origin and this integral is evidently zero. We are consequently justified in regarding the singularity as a doublet and in fact

* For a fuller discussion see Larmor, *Phil. Mag.* (5), Vol. 44 (1897), p. 503; *Aether and Matter*, Chap. XIV, where it is shown that energy is radiated from a moving charge only when the velocity of the charge alters in either magnitude or direction.

as a simple electric doublet of varying moment as is indicated by the way in which the electric and magnetic forces become infinite*. The axis of the doublet is along the axis of z.

The electric lines of force due to a vibrating electric doublet have been drawn by Hertz† for various stages of the motion. The general character of the lines of force is indicated in Fig. 1.

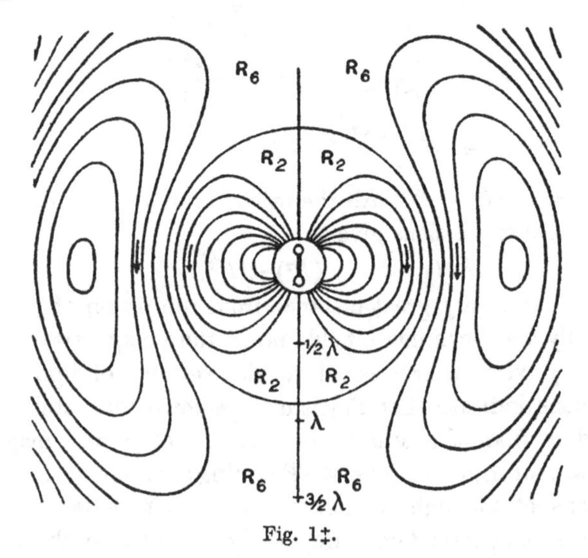

Fig. 1‡.

It will be noticed that the lines are all at right angles to a plane perpendicular to the axis of the doublet. M. Abraham§ has used a Hertzian doublet to obtain a model of the electromagnetic field produced by the oscillations in a vertical antenna, the plane just mentioned being supposed to represent the earth which is regarded as a perfect conductor. Zenneck‖ has, however, pointed out that when the imperfect conductivity of the earth is taken into account the circumstances of the

* See § 42.

† *Ann. d. Phys.* Vol. 36 (1888), p. 1. The case of damped vibrations is considered by K. Pearson and A. Lee, *loc. cit.*

‡ I am indebted to the Macmillan Company and A. Gray, Esq., for permission to reproduce this diagram.

§ *Phys. Zeitschr.* Vol. 2 (1901), p. 329; *Theorie der Elektrizität*, Vol. 2, § 34; *Encyklop. d. Math. Wiss.* Band 5, § 18.

‖ *Ann. d. Phys.* Vol. 23 (1907), p. 846; *Phys. Zeitschr.* Vol. 9 (1908), p. 50; *Ibid.* p. 553.

propagation are somewhat different. The spreading of electromagnetic waves over the earth's surface has been investigated thoroughly by A. Sommerfeld* and his pupil H. v. Hoerschlemann†, and their results seem to indicate that the imperfect conductivity of the earth is an important factor in directing electric waves and in enabling their effects to be detected at great distances. The ionisation of the air by sunlight is also an important factor, as has been pointed out by J. J. Thomson, W. H. Eccles‡ and J. A. Fleming§. Marconi's experiments have indicated that the circumstances of propagation are not yet thoroughly understood. No good reason has been given to explain why communications by means of electric waves can be made more easily when the receiving station is in a north or south direction than when the direction is east or west. The curious contrasts in the results obtained with waves of different frequencies in day and night communications are also unexplained‖.

The use of the vector Π instead of the scalar S was recommended by Abraham¶. Von Hoerschlemann has obtained in this way a model of Marconi's bent antenna which gives a directed effect to the radiation. A number of arrangements of Hertzian doublets that can be used to imitate the action of antennae have been described by Fleming**, Larmor††, Sommerfeld and Macdonald‡‡.

In the theory of FitzGerald's magnetic oscillator§§ we have

$$\Pi = 0, \quad \Gamma = (0, 0, N),$$

N being Euler's wave-function. Whittaker's solution is obtained by adding the solutions of Hertz and FitzGerald.

* *Ann. d. Phys.* Vol. 28 (1909), p. 665.

† *Jahrb. d. draht. Teleg.* Vol. 5 (1912).

‡ *Proc. Roy. Soc.* A, Vol. 87, p. 79.

§ *British Association Reports*, Dundee (1912). See also O. J. Lodge, *Phil. Mag.* Vol. 25 (1913), p. 775.

‖ See Marconi's address to the Royal Institution, June, 1913.

¶ *Theorie der Elektrizität*, Vol. 2, Ch. I. See also Righi, *loc. cit.*

** *Proc. Roy. Soc.* A, Vol. 78, p. 1.

†† *Ibid.* in a footnote to Fleming's paper.

‡‡ *Proc. Roy. Soc.* A, Vol. 81, p. 394.

§§ *Trans. Roy. Dublin Soc.* Vol. 3 (1883); *Scientific Writings*, p. 122.

§ 5. Second solution of the fundamental equations.

It is easy to see that equations (2) will be satisfied if we can find two functions (α, β) such that

$$\left.\begin{aligned} M_x &= \frac{\partial(\alpha, \beta)}{\partial(y, z)} \equiv \pm \frac{i}{c} \frac{\partial(\alpha, \beta)}{\partial(x, t)} \\ M_y &= \frac{\partial(\alpha, \beta)}{\partial(z, x)} \equiv \pm \frac{i}{c} \frac{\partial(\alpha, \beta)}{\partial(y, t)} \\ M_z &= \frac{\partial(\alpha, \beta)}{\partial(x, y)} \equiv \pm \frac{i}{c} \frac{\partial(\alpha, \beta)}{\partial(z, t)} \end{aligned}\right\} \quad \ldots\ldots\ldots\ldots\ldots(10).$$

An electromagnetic field that is specified in this way is necessarily a self-conjugate field, for if we multiply together the two expressions for M_x and do the same for M_y, M_z, we find that $M^2 = 0$. A particular pair of functions α, β is obtained by putting

$$\alpha = x\cos\theta + y\sin\theta \mp iz, \quad \beta = x\sin\theta - y\cos\theta - ct \;\ldots\ldots(11),$$

where θ is an arbitrary constant. To generalise this field we multiply the expressions for M_x, M_y, M_z by an arbitrary function* of α, β, θ and integrate with regard to θ; we thus obtain a very general electromagnetic field in which

$$\left.\begin{aligned} M_x &= \mp i \int_0^{2\pi} f(\alpha, \beta, \theta)\cos\theta\, d\theta \\ M_y &= \mp i \int_0^{2\pi} f(\alpha, \beta, \theta)\sin\theta\, d\theta \\ M_z &= - \int_0^{2\pi} f(\alpha, \beta, \theta)\, d\theta \end{aligned}\right\} \quad \ldots\ldots\ldots\ldots(12).$$

The components of the electric and magnetic forces are obtained by equating the ambiguous and unambiguous parts in these equations; it is easy to verify that they are all wave-functions.

It should be remarked that these definite integrals may give a representation of the electromagnetic field, required for the solution of a problem, only in a certain limited domain of

* When we speak of an arbitrary function it must be understood that the function may be subject to certain limitations which render the integration and differentiation under the integral sign intelligible operations.

the variables x, y, z, t; the integrals may in fact represent discontinuous functions.

The limits of integration could have been taken to be any other constants instead of 0 and 2π; they can also be taken to be functions of x, y, z, t of the type ω, where ω is defined by an equation of the form

$$x \sin \omega - y \cos \omega - ct = F(\omega),$$

F being an arbitrary function.

A suitable pair of functions α, β is also obtained by putting

$$\alpha = \frac{x \mp iy}{z + r}, \quad \beta = r - ct \quad \text{.................(13)},$$

and in this case Poynting's vector is along the radius from the origin. A more general type of electromagnetic field in which this is true is obtained by multiplying the above expressions for the components of M by an arbitrary function of α and β.

Other pairs of functions α, β of a very general nature are obtained in Chap. VIII. It should be remarked that in all cases the functions (α, β) are of such a nature that if $F(\alpha, \beta)$ is an arbitrary function of α and β, F satisfies the partial differential equation

$$\left(\frac{\partial F}{\partial x}\right)^2 + \left(\frac{\partial F}{\partial y}\right)^2 + \left(\frac{\partial F}{\partial z}\right)^2 = \frac{1}{c^2}\left(\frac{\partial F}{\partial t}\right)^2 \quad \text{.........(14)},$$

which is of fundamental importance in geometrical optics* and may be called *Hamilton's equation.* It is found that this equation is also satisfied in many cases by the functions of x, y, z, t which are the limits of a definite integral representing a wave-function, when the function under the integral sign is a wave-function for all values of the parameter with regard to which we are integrating. Thus the function ω just defined and the function $t - \frac{1}{c}(r + r_0)$ which will be used later are solutions of this equation.

* For another connection between this equation and the electromagnetic equations see A. Sommerfeld and J. Runge, "Grundlagen der geometrischen Optik," *Ann. d. Phys.* Vol. 35 (1911), p. 277.

§ 6. The fundamental equations for a material medium.

For a material medium which is stationary relative to the axes of coordinates, the equations (1) must be replaced by the more general equations*

$$\left.\begin{aligned} \operatorname{rot} H &= \frac{1}{c}\left(J + \frac{\partial D}{\partial t}\right), \quad &\operatorname{div} D = \rho \\ \operatorname{rot} E &= -\frac{1}{c}\frac{\partial B}{\partial t}, \quad &\operatorname{div} H = 0 \end{aligned}\right\} \quad \ldots\ldots\ldots(15),$$

where D is the electric displacement, E the electric force or field strength, H the magnetic force and B the magnetic induction. The quantity $J + \frac{\partial D}{\partial t}$ represents the total current which is made up of a conduction-current C, a displacement-current $\frac{\partial D}{\partial t}$ and a convection-current ρv, ρ being the volume density of electricity.

Various notations have been used for the different vectors of an electromagnetic field. Most English writers use (a, b, c) for the components of the magnetic induction, (α, β, γ) for those of the magnetic force, (f, g, h) for the components of the electric displacement and (P, Q, R) or (X, Y, Z) for those of the electromotive intensity or electric force†. This is not to be confused with the mechanical force F of electromagnetic origin, whose components are sometimes denoted by (X, Y, Z).

* Lorentz (1892—1895) and Larmor (1895) have derived these equations and a corresponding set of equations for moving bodies by a process of averaging, starting from the fundamental equations of the theory of electrons in which we have $B=H$, $D=E$, $J=\rho v$. Cf. H. A. Lorentz, *Akad. van Wetenschappen te Amsterdam* (1902), p. 305; *Encykl. d. Math. Wiss.* Bd. 5, § 14, pp. 200—210. This method of averaging has been developed so as to give results in accordance with the Theory of Relativity by M. Born, *Math. Ann.* Bd. 68 (1910) and E. Cunningham, *Proc. London Math. Soc.* Ser. 2, Vol. 10 (1911), p. 116. The Born-Minkowski equations differ slightly from those of Lorentz and indicate the existence of an electrostatic field due to the motion of a magnetised body.

It has been realised by the foregoing writers and others that the principle of relativity alone is not sufficient to determine a complete set of equations for moving bodies, a theory of the constitution of matter is needed. Cf. H. R. Hassé, *Phil. Mag.* Jan. (1914).

† Clerk Maxwell, *Electricity and Magnetism*, 3rd edition (1892), Vol. 2, p. 257.

In any material medium there are certain constitutive relations connecting the vectors D, E, B, H, J. In moving media, crystalline media and ferromagnetic bodies the relations are rather complicated, but for an isotropic medium in which $\rho = 0$ the relations can be represented to a good degree of approximation by the simple equations

$$D = \epsilon E, \quad J = \sigma E, \quad B = \mu H \quad \ldots\ldots\ldots\ldots\ldots(16),$$

where ϵ, σ, μ are scalar quantities which are generally regarded as constants; they will be regarded in fact as the optical constants of the medium. The quantity σ is called the *conductivity*, μ the *permeability*, and ϵ the *dielectric-inductive capacity*.

The units that are used here are the so-called modified units*, in which Heaviside's suggestion of eliminating a factor 4π has been adopted. We can pass to electrostatic units or electromagnetic units by replacing our quantities E, H, etc. by αE, βH, etc., where α, β are certain factors which are given in the following table:

	e, D, J	E	B	H	m
Electrostatic system	$\sqrt{4\pi}$	$\frac{1}{\sqrt{4\pi}}$	$\frac{c}{\sqrt{4\pi}}$	$\frac{1}{c\sqrt{4\pi}}$	$c\sqrt{4\pi}$
Electromagnetic system	$c\sqrt{4\pi}$	$\frac{1}{c\sqrt{4\pi}}$	$\frac{1}{\sqrt{4\pi}}$	$\frac{1}{\sqrt{4\pi}}$	$\sqrt{4\pi}$

We use e here to denote a quantity of electricity, and m a quantity of magnetism.

§ 7. The energy equation for a material medium.

If we use Σ as before to denote the vector whose components are of type $c(E_y H_z - E_z H_y)$, we find that

$$\frac{\partial \Sigma_x}{\partial x} + \frac{\partial \Sigma_y}{\partial y} + \frac{\partial \Sigma_z}{\partial z} + \frac{\partial}{\partial t}\frac{\epsilon}{2}(E^2) + H_x \frac{\partial B_x}{\partial t} + H_y \frac{\partial B_y}{\partial t} + H_z \frac{\partial B_z}{\partial t} + \sigma(E^2) = 0.$$

* Cf. H. A. Lorentz, *Encyklopädie der Math. Wiss.* Bd. 5, § 13, pp. 83—87.

If μ is a constant and $\frac{1}{2}\epsilon E^2 + \frac{1}{2}\mu H^2$ be regarded as the energy per unit volume, the change in the distribution of the energy can be described by means of a flow σ and a loss per unit volume of magnitude σE^2 due to the transformation of electric energy into heat (Joule's heat)*. If B does not depend on the instantaneous value of H so that μ is not a constant there is a loss of energy due to hysteresis. B may depend upon H alone but not be a single-valued function of H, consequently in a cycle of changes $\int(H.dB)$ is not zero and may be taken as the heat per unit volume developed during the description of the cycle. Notice that

$$\int(H.dB) = -\int(B.dH) \text{ round a cycle,}$$

and is always positive since the value of B for a given value of H is greater when H is increasing. The experimental analysis and the accompanying theory are due to E. Warburg† and independently in much greater development to J. A. Ewing‡ by whom the name hysteresis was applied to such phenomena.

§ 8. **Solution of the fundamental equations for a material medium.**

Let us assume that σ, μ, ϵ are constants and that E, H are the real parts of expressions of the form $Ae^{-i\omega t}$, where A is a complex quantity independent of t. Then if we write

$$k^2 = \frac{\epsilon\mu\omega^2 + i\mu\omega\sigma}{c^2}, \quad \nu = -\frac{kc}{\mu\omega}, \quad M = H \pm i\nu E \text{ ......(17),}$$

and regard E, H now as the complex vectors of which they were formerly the real parts, the differential equations to be satisfied by M are

$$\text{rot } M = \pm kM, \quad \text{div } M = 0 \text{ ....(18).}$$

These may be solved by putting

$$M = \text{rot } \Pi \pm \frac{1}{k} \text{grad div } \Pi \pm k\Pi \text{ ....(19),}$$

where Π is a solution of the equation $\Delta u + k^2 u = 0$, and may be of the form $U \pm iV$.

* For a fuller discussion see R. Gans, *Einführung in die Theorie der Magnetismus*; H. A. Lorentz, *Encykl. d. Math. Wiss.* Bd. 5, § 14, p. 240, Heft 1 (1903); Heaviside, *Electrical Papers*, Vol. 1, pp. 437—450.

† *Ann. Phys. Chem.* (3), Vol. 13 (1881), p. 141.

‡ *Phil. Trans.* A, Vol. 176 (1885), pp. 523—640; *Proc. Roy. Soc.* Vol. 34 (1883), p. 39; *Phil. Mag.* Vol. 16 (1883), p. 381; *Magnetic Induction in Iron and other Metals*, London (1892).

Problems which deal with the effect of small obstacles upon light or electric waves require the taking into account of the properties of the material of which the obstacle is composed: it is only for long waves and conducting media that the obstacle can be treated as a perfect reflector. This is illustrated by the work of Maxwell Garnett* and Mie† on the optical properties of colloidal suspensions of metals and in the work of Sommerfeld to which we have already referred.

Unfortunately, however, the analytical difficulties are very great when imperfect conductivity is taken into account‡. The simple-looking problem of the reflection of the disturbance produced by a moving charge, when the obstacle is an infinite plane sheet of metal or other conducting substance, has not yet been solved accurately§ and there are many similar problems that have completely baffled mathematicians.

Much more progress has been made with problems dealing with perfect reflectors. These problems are to some extent ideal but some of the characteristics of actual physical problems are often preserved‖. Apart from this, such problems are of considerable mathematical interest, and have been studied by some writers simply on this account.

§ 9. Boundary-Conditions.

The conditions to be satisfied at a surface separating two different media are obtained by integrating equations (15) across a thin layer of transition¶. Taking the axis of z along the

* *Phil. Trans.* A, Vol. 203 (1904), p. 385; Vol. 205 (1905), p. 237.

† *Ann. d. Phys.* Vol. 25 (1908), p. 377.

‡ An important solution of the equations has been given by M. Brillouin, "Propagation dans les milieux conducteurs," *Comptes Rendus*, t. 136 (1903), pp. 667, 746. (See Ex. 20, Ch. IX.) For other references see Ex. 5, p. 23.

§ An approximate solution was suggested by Maxwell and has been developed by Larmor, *Proc. London Math. Soc.*, Ser. 2, Vol. 8, p. 1. An accurate solution for the case in which the sheet is treated as infinitely thin and the charge moves with uniform velocity parallel to the plane has been given by G. Picciati, *Rom. Acc. Linc. Rend.* (5), 11_2 (1902), p. 221.

‖ A metal behaves practically as a perfect conductor to electric waves when the oscillations are rapid but slow compared with waves of light. Cf. J. Larmor, "Electric vibrations in condensing systems," *Proc. London Math. Soc.* Ser. 1, Vol. 26, p. 119.

¶ Rayleigh, *Scientific Papers*, Vol. 1; *Phil. Mag.* Vol. 12 (1881), p. 81; H. Hertz, *Electric Waves*, pp. 207, 238; Larmor, *Phil. Trans.* (1895),

normal to the surface we shall assume that E, H, D, B and their derivatives with regard to x, y, t are finite within the layer and that the conductivity σ is also finite. The equations

$$\frac{\partial E_y}{\partial z} - \frac{\partial E_z}{\partial y} = \frac{1}{c}\frac{\partial B_x}{\partial t}, \quad \frac{\partial H_z}{\partial y} - \frac{\partial H_y}{\partial z} = \frac{1}{c}\left(\sigma E_x + \frac{\partial D_x}{\partial t}\right),$$

then show that $\frac{\partial E_y}{\partial z}$, $\frac{\partial H_y}{\partial z}$ and $J_x + \frac{\partial D_x}{\partial t}$ are finite and so their integrals with regard to z across a thin layer of thickness θ are less than $a\theta$ where a is a finite positive quantity independent of θ. This means that the tangential components of the electric force, magnetic force and electric current are continuous in crossing the surface.

Again, the equation

$$\frac{\partial B_x}{\partial x} + \frac{\partial B_y}{\partial y} + \frac{\partial B_z}{\partial z} = 0$$

shows that $\frac{\partial B_z}{\partial z}$ is finite and so the normal component of the magnetic induction is continuous. This result may, however, be regarded as a consequence of the previous one.

The normal component of the electric displacement may be discontinuous, for the equation $\text{div}\, D = \rho$ gives

$$d_z = \int_0^\theta \frac{\partial D_z}{\partial z} dz = \int_0^\theta \rho dz + \text{terms of order } \theta,$$

where d_z is the discontinuity in the electric displacement. Hence if $\bar{\sigma}$ is the surface-charge of electricity per unit area we have $d_z = \bar{\sigma}$. When the media are both conducting we have $\bar{\sigma} = 0$ and the normal component of the electric displacement is continuous.

It is known that in the case of a good conducting body rapidly alternating currents are confined within a very thin layer close to the surface*. In the ideal case of a perfect conductor or perfect reflector the field-vectors are zero within

p. 733; H. M. Macdonald, *Electric Waves*, p. 14; H. A. Lorentz, *Zeitschrift f. Math. u. Phys.* Bd. 22 (1877).

* Cf. H. Lamb, *Phil. Trans.* A (1883); O. Heaviside, *Electrical Papers*, Vol. 2, p. 168; J. J. Thomson, *Recent Researches*, p. 281. For some recent work on the subject see a paper by E. F. Northrup and J. R. Carson, *Journ. of the Franklin Institute*, Feb. (1914), p. 125. The results of many other researches on the skin-effect and alternating current resistance are given in J. A. Fleming's *The principles of Electric Wave Telegraphy*.

the body of the conductor. This case is characterised by $\sigma = \infty$ and the tangential components of the magnetic force are no longer continuous as a point moves across the surface, we have in fact if h is the discontinuity in the magnetic force

$$h_y = \int_0^\theta \frac{\partial H_y}{\partial z} dz = -\frac{1}{c}\int_0^\theta \sigma E_x dz + \text{terms of order } \theta.$$

Hence if K is the surface-current and h the discontinuity of the magnetic force, we have

$$h_y = -\frac{1}{c} K_x, \quad h_x = \frac{1}{c} K_y.$$

At the surface of a perfect conductor the tangential components of the electric force must vanish and as a consequence of this we can say that the normal component of the magnetic induction must also vanish. If the medium outside the conductor is free aether the surface-conditions are simplified on account of the relations $D = E$, $B = H$.

In the case of a very thin conducting sheet it is convenient to treat the thickness of the sheet as negligible and regard the tangential components of the magnetic force as discontinuous when a point moves along the normal from one side of the sheet to the other.

The boundary-condition is then*

$$-h_y = \frac{\bar{\sigma}}{c} E_x, \quad h_x = \frac{\bar{\sigma}}{c} E_y,$$

where $\bar{\sigma} = \int_0^\theta \sigma dz$ is the conductivity of the sheet.

If we wish to extend the idea of Green's equivalent layer to electrodynamics we must consider electromagnetic fields in free aether with surfaces at which the tangential components of the electric and magnetic forces are discontinuous; this requires an electric current sheet and a magnetic current sheet on the

* Cf. T. Levi-Cività, *Rend. Lincei* (5), 11_2 (1902), p. 75. These conditions are used by Picciati in his solution of the problem of an electric charge moving parallel to a conducting sheet. In some previous papers, *Rend. Lincei* (5), 11_1 (1902), pp. 163, 191, 228, Levi-Cività had used the electromagnetic potentials to determine the effect of a conducting sheet on an alternating current flowing along a straight wire parallel to the sheet; the boundary-conditions are then determined by the discontinuities of the potentials due to the induced current.

surface*. For a complete generalisation we ought to consider the cases when the normal components are also discontinuous and when the surface is in motion. The last circumstance alters matters to some extent and must now be discussed.

In the case of electric waves in free aether, the vectors E and H may be discontinuous at a wave-boundary. If this can be regarded as the limit of a thin layer of transition within which equations (1) are satisfied and the vectors E, H are finite†, the values of these vectors on the two sides of the boundary must satisfy certain conditions which may be found as follows.

Let the equation of the moving boundary be expressed in the form

$$t = f(x, y, z) \qquad \text{.......................(20).}$$

If now we apply Green's theorem to the integral

$$\iint \Big[\left(M_x \mp icM_y \frac{\partial t}{\partial z} \pm icM_z \frac{\partial t}{\partial y} \right) dy dz + \left(M_y \mp icM_z \frac{\partial t}{\partial x} \pm icM_x \frac{\partial t}{\partial z} \right) dz dx + \left(M_z \mp icM_x \frac{\partial t}{\partial y} \pm icM_y \frac{\partial t}{\partial x} \right) dx dy \Big] \qquad \text{......(21),}$$

which is supposed to be taken over a closed surface, we find that it vanishes on account of the equations (2) provided t is supposed to be expressed in terms of x, y, z according to some definite law which we shall take to be that expressed by (20).

We now apply this theorem to a disc-shaped surface whose two faces very nearly coincide. We shall suppose that on one side of the disc the vector M represents the field of the advancing waves and that on the other side it represents the field obtaining just before the arrival of the waves. We shall also suppose

* Cf. J. Larmor, *Proc. London Math. Soc.* (2), Vol. 1 (1903), p. 11; H. M. Macdonald, *Electric Waves*, p. 16; *Proc. London Math. Soc.* (2), Vol. 10 (1911), p. 91.

† The idea is practically due to Stokes, *Math. and Phys. Papers*, Vol. 2, p. 275, but was not worked out in detail. The different possible types of discontinuity are discussed with some care by Love. The case in which E and H are continuous but some of their derivatives are discontinuous at the moving boundary may be discussed more simply by analysis analogous to that given in Hadamard's *Leçons sur la Propagation des Ondes*, Paris (1903), Ch. 2. See also Ex. 2, p. 23, and the references to Duhem and Silberstein on the next page.

that the derivatives of M are finite or behave in such a way that an application of Green's theorem is justifiable. Now let $\bar{M}$ be the discontinuity of M, i.e. the difference in the values of M at two neighbouring points on opposite sides of the wave-boundary. Then when the two faces of the disc coincide we find that a certain surface-integral over one face of the disc is zero. The surface-integral is of the same type as (21) except that $\bar{M}$ is written in place of M. Since the face of the disc can be chosen arbitrarily the integrand must vanish and so we obtain three equations of the type*

$$\bar{M}_x \mp ic\,\bar{M}_y \frac{\partial t}{\partial z} \pm ic\,\bar{M}_z \frac{\partial t}{\partial y} = 0 \quad \text{...........(22)}.$$

These equations give

$$\bar{M}^2 = 0 \quad \text{and} \quad \left(\frac{\partial t}{\partial x}\right)^2 + \left(\frac{\partial t}{\partial y}\right)^2 + \left(\frac{\partial t}{\partial z}\right)^2 = \frac{1}{c^2}.$$

Hence the wave-front advances with the velocity of light and the difference between the two electromagnetic fields at the wave-boundary behaves as a self-conjugate field in which Poynting's vector is along the normal to the moving boundary.

If the equation of the wave-boundary be expressed in the form

$$F(x, y, z, t) = 0,$$

we find on calculating the values of

$$\frac{\partial t}{\partial x},\ \frac{\partial t}{\partial y},\ \frac{\partial t}{\partial z},$$

that

$$\left(\frac{\partial F}{\partial x}\right)^2 + \left(\frac{\partial F}{\partial y}\right)^2 + \left(\frac{\partial F}{\partial z}\right)^2 = \frac{1}{c^2}\left(\frac{\partial F}{\partial t}\right)^2 \quad \text{......(14)}.$$

This is the differential equation of the characteristics, it expresses that the moving boundary moves normally to itself with the velocity of light. According to the theory of Stokes†

* Equations equivalent to these are obtained in a different manner by O. Heaviside, *Electrical Papers*, Vol. 2, p. 405; A. E. H. Love, *Proc. London Math. Soc.* Ser. 2, Vol. 1 (1903), p. 37; L. Silberstein, *Ann. d. Physik*, Vol. 26 (1908), p. 751; P. Duhem, *Comptes Rendus*, t. 131 (1900), p. 1171.

† *Proc. Camb. Phil. Soc.* (1896); *Manchester Memoirs* (1897); *Math. and Phys. Papers*, Vol. 4.

and Wiechert*, Röntgen rays consist of pulses travelling through the aether, the energy in a pulse being confined within a thin shell. The above theory indicates that the front and rear surfaces of the shell move forward with the velocity of light.

A slight modification of the preceding method can be used to find the conditions to be satisfied at a moving surface which is the boundary between two different media. If we write $L = B \pm iD$, $N = E \mp iH$ and assume that the surface-charge and surface-current can be neglected, the six boundary-conditions can be expressed by saying that the three quantities of type

$$L_x - c\frac{\partial t}{\partial y}N_z + c\frac{\partial t}{\partial z}N_y$$

must be continuous as the boundary is crossed. The equation of the moving boundary is expressed as before in the form (20). When the moving boundary is the surface of a perfect conductor or perfect reflector, the boundary-conditions are simply that the three quantities of type

$$B_x - c\frac{\partial t}{\partial y}E_z + c\frac{\partial t}{\partial z}E_y$$

should vanish†.

In the last two cases the boundary-conditions do not imply that the boundary moves normally to itself with the velocity of light; in fact, the motion of the boundary can be quite arbitrary.

EXAMPLES.

1. The surface of discontinuity is the sphere $r=ct$ and the electromagnetic field within this surface is expressed by the equations

$$(E_x, E_y, E_z)=c\left(\frac{\partial^2\Pi}{\partial x\partial z}, \quad \frac{\partial^2\Pi}{\partial y\partial z}, \quad -\frac{\partial^2\Pi}{\partial x^2}-\frac{\partial^2\Pi}{\partial y^2}\right),$$

$$(H_x, H_y, H_z)=\left(\frac{\partial^2\Pi}{\partial y\partial t}, \quad -\frac{\partial^2\Pi}{\partial x\partial t}, \quad 0\right),$$

where $$\Pi=Ar^{-1}e^{-q(ct-r)}\sin p\,(ct-r+\epsilon).$$

* *Abh. d. Phys.-ökon. Ges. zu Königsberg*, 1 Pr. (1896), p. 1; *Ann. Phys. Chem.* Bd. 59 (1896), p. 283. See also J. J. Thomson, *Phil. Mag.* (5), Vol. 45 (1898), p. 172.

† These conditions are equivalent to the condition that the mechanical force, which would act on a charge moving with the normal velocity of the surface, must be along the normal to the surface. Cf. Heaviside, *Electrical Papers*, Vol. 2, p. 514; *Electromagnetic Theory*, Vol. 1, p. 273; Hertz, *Electric Waves*, p. 257.

The constants p and q being known determine the constants A and ϵ in order that the field outside the surface $r=ct$ may be the electrostatic field for which the potential Φ is $-\frac{\partial}{\partial z}\left(\frac{1}{r}\right)$.

(Cambr. Math. Tripos, Part II, 1904.)

2. If the vectors E and H are known for all points (x, y, z, t) of a moving surface

$$F(x, y, z, t)=0$$

the values of all the derivatives of E and H, and consequently values of E and H at points not on the moving surface can generally be found provided F does not satisfy the differential equation

$$\left(\frac{\partial F}{\partial x}\right)^2+\left(\frac{\partial F}{\partial y}\right)^2+\left(\frac{\partial F}{\partial z}\right)^2=\frac{1}{c^2}\left(\frac{\partial F}{\partial t}\right)^2.$$

(Havelock, *Proc. London Math. Soc.* Ser. 2, Vol. 2, p. 297.)

3. An electromagnetic field is conjugate to an electrostatic field. Prove that the flow of energy in the electromagnetic field takes place along the lines of electric force in the electrostatic field.

4. Let the line OC of length c be drawn in the direction of Poynting's vector at each space-time point O of a self-conjugate electromagnetic field and let OV represent a velocity v associated with the point O. Prove that if this electromagnetic field is conjugate to another field (E, H) in which cH is the vector product of v and E, the direction of E is parallel to VC.

5. Prove that when σ, ϵ, μ are constants, the vectors E, H of § 6 satisfy the differential equations

$$\Delta E-\frac{\epsilon\mu}{c^2}\frac{\partial^2 E}{\partial t^2}-\frac{\sigma\mu}{c^2}\frac{\partial E}{\partial t}=\frac{1}{\epsilon}\operatorname{grad}\rho,$$

$$\Delta H-\frac{\epsilon\mu}{c^2}\frac{\partial^2 H}{\partial t^2}-\frac{\sigma\mu}{c^2}\frac{\partial H}{\partial t}=0. \qquad \text{(Maxwell.)}$$

Solutions of these equations for the case in which $\rho=0$ have been given by

O. Heaviside, *Phil. Mag.*, Jan. (1889), p. 30 ; *Electrical Papers*, Vol. 2, p. 478.

H. Poincaré, *Comptes Rendus* (1893), p. 1030 ; *Théorie analytique de la propagation de la chaleur*, Ch. 8.

J. Boussinesq, *Comptes Rendus* (1894), pp. 162—223 ; *Théorie analytique de la chaleur*, t. 2 (1903), p. 538.

Kr. Birkeland, *Archives des sciences physiques*, Geneva (1895), p. 5.

O. Tedone, *Rend. Lincei*, Mar. 31st (1913), Jan. 18th (1914).

M. I. Pupin, *Trans. Amer. Math. Soc.*, Vol. 1 (1900), p. 259.

6. Prove that the function

$$u=\frac{1}{\sqrt{t-x}}e^{-\frac{\sigma}{2}[t\pm\sqrt{t^2-x^2}]}$$

satisfies the equation

$$\frac{\partial^2 u}{\partial x^2} = \frac{\partial^2 u}{\partial t^2} + \sigma \frac{\partial u}{\partial t}.$$

7. If u satisfies the partial differential equation

$$\left(\frac{\partial u}{\partial x}\right)^2 + \left(\frac{\partial u}{\partial y}\right)^2 + \left(\frac{\partial u}{\partial z}\right)^2 = \frac{1}{c^2}\left(\frac{\partial u}{\partial t}\right)^2,$$

and we make the transformation

$$x_0 = x + f\frac{\partial u}{\partial x}, \qquad z_0 = z + f\frac{\partial u}{\partial z},$$

$$y_0 = y + f\frac{\partial u}{\partial y}, \qquad t_0 = t - \frac{f}{c^2}\frac{\partial u}{\partial t},$$

where f is an arbitrary function, then u also satisfies the partial differential equation

$$\left(\frac{\partial u}{\partial x_0}\right)^2 + \left(\frac{\partial u}{\partial y_0}\right)^2 + \left(\frac{\partial u}{\partial z_0}\right)^2 = \frac{1}{c^2}\left(\frac{\partial u}{\partial t_0}\right)^2.$$

8. Plane electromagnetic waves fall on the convex surface of an infinite paraboloid of revolution $x = a - r$, whose surface is a perfect reflector. If the incident waves are given by expressions of the type

$$H_x + iE_x = f(\alpha, \beta)\frac{\partial(\alpha, \beta)}{\partial(y, z)} = \frac{i}{c} f(\alpha, \beta)\frac{\partial(\alpha, \beta)}{\partial(x, t)},$$

where $\alpha = y + iz$, $\beta = x + ct$, the boundary-conditions at the surface of the paraboloid may be satisfied by subtracting from the primary field a secondary field represented by expressions of a similar type but with

$$\alpha = a\frac{y - iz}{x + r}, \qquad \beta = a - r + ct.$$

If waves represented by the above expressions with $\alpha = y - iz$, $\beta = x - ct$, fall on the concave surface of the paraboloid the boundary-conditions at the surface of the paraboloid may be satisfied by supposing that the secondary disturbance is of the form $-M' - M'' + M'''$, where the fields M', M'', M''' are represented by expressions of the above type, where α, β have the values

$$\alpha' = y + iz, \qquad \beta' = 2a - x - ct,$$

$$\alpha'' = a\frac{y + iz}{x + r}, \qquad \beta'' = a - r - ct,$$

$$\alpha''' = a\frac{y - iz}{x + r}, \qquad \beta''' = a + r - ct,$$

respectively. With these suppositions the forces are finite at the focus, when f is independent of α.

CHAPTER II

GENERAL SURVEY OF THE DIFFERENT METHODS OF SOLVING THE WAVE-EQUATION

§ 10. The object to be attained.

It has been shown in Chapter I that the solution of Maxwell's equations can be made to depend upon the solution of a single partial differential equation which is either the wave-equation $\Omega u = 0$ or the equation $\Delta u + k^2 u = 0$ which is satisfied by wave-functions of the form $u = e^{\pm ikct} f(x, y, z)$. The properties of functions satisfying these equations must accordingly be studied at some length. It is desirable, also, that all types of such functions should be studied and not merely those which admit readily of application to physical problems. If certain solutions of the fundamental equations must be rejected in the treatment of the boundary problems of mathematical physics, a knowledge of their behaviour is at any rate useful as it gives a clear indication of the reason why such solutions must be rejected. There is, however, another reason why the scope of the inquiry should not be restricted. The theory of wave-functions forms a natural extension of the theory of functions of a complex variable* and may consequently lead to results of great value for the general theory of functions.

* This point of view is adopted, for instance, by Volterra, *Rend. Lincei* (4), III_2, pp. 225–330, 274–287 (1887); IV_1, pp. 107–115, 196–202 (1889); V_1, pp. 158–165, 291–299, 599–611, 630–640 (1889); *Rend. Palermo*, Vol. 3, pp. 260–272. See also Appell, *Acta Math.* t. 4, p. 313 (1884); Painlevé, *Toulouse Ann.*, t. 2B (1888); Bôcher, *Bull. Amer. Math. Soc.* Vol. 9, p. 455 (1903). The theory of functions of two complex variables is closely connected with the theory of wave-functions. Cf. H. Poincaré, *Acta Math.* Vol. 2 (1883); Vol. 22 (1898); H. F. Baker, *Camb. Phil. Trans.* Vol. 18 (1899), p. 431; *Proc. London Math. Soc.* Ser. 2, Vol. 1 (1903), p. 14.

We shall now describe briefly some of the principal methods of solving the wave-equation.

§ 11. Reduction to ordinary differential equations. The aim of this method is to determine elementary solutions of the form

$$u = f_1(\alpha) f_2(\beta) f_3(\gamma) f_4(\delta) \quad \text{..............(23)}$$

where f_1, f_2, f_3, f_4 are particular functions of their arguments and $\alpha, \beta, \gamma, \delta$ are particular functions of x, y, z, t. This method was used by D. Bernoulli in 1732 in the treatment of the vibrations of a hanging chain, the partial differential equation being however in this case different.

The general theory of elementary solutions is due to Lamé* who transformed Laplace's equation into curvilinear coordinates. For a historical account of the development of the theory we may refer to Prof. Bôcher's book *Die Reihenentwickelungen der Potentialtheorie*, Leipzig (1894) and to Byerly's *Fourier Series and Spherical Harmonics*.

A simple elementary solution of the wave-equation is obtained by putting $\alpha = x$, $\beta = y$, $\gamma = z$, $\delta = t$; we can then take

$$u = e^{lx + my + nz \pm pt} \quad \text{.................(24)}$$

where the constants l, m, n, p satisfy the relation

$$p^2 = c^2 (l^2 + m^2 + n^2) \quad \text{.................(25)}$$

and can be either real or complex quantities.

When p is a purely imaginary quantity and l, m, n unrestricted, the solution is periodic and more general periodic solutions may be derived from this one by summation, l, m, n being regarded as variable parameters subject to the relation (25). When l, m, n are purely imaginary the solution (24) is appropriate for the representation of plane waves of monochromatic light, the intensity and phase of which are the same at all points of any plane perpendicular to the direction of propagation.

* *Liouville's Journal*, t. 2, pp. 147–183; *Leçons sur les coordonnées curvilignes*, Paris (1859). See also E. Mathieu, *Cours de physique mathématique* (1873).

Taking $m = n = 0$, $l = -\frac{p}{c} = \nu$ so that the axis of x is in the direction of propagation, we may write

$$E_x = E_z = H_x = H_y = 0,\ E_y = a \cos \nu (x - ct),$$
$$H_z = a \cos \nu (x - ct) \ldots\ldots\ldots\ldots\ldots\ldots(26)$$

where a is a constant. These waves will be said to be linearly polarised in a direction parallel to the axis of y and will be called homogeneous because E and H do not depend on y and z. It will be noticed that the constant c represents the velocity of propagation of a phase of the disturbance.

A wave-function of the type

$$\Omega = \sin \nu x \cos \nu ct$$

is appropriate for the representation of standing waves. Expressions for E and H may be written down by analogy with the above. To obtain a representation of plane waves in a conducting medium, we must use a solution of

$$\frac{d^2u}{dx^2} + k^2u = 0,$$

where k^2 has the complex value given in § 8. Putting $V = 0$, $U = (0, 0, e^{ikx - i\omega t})$ we find that

$$E_x = E_y = H_x = H_z = 0,\ E_y = \frac{i\mu\omega}{c} e^{ikx - i\omega t},$$
$$H_z = ik\, e^{ikx - i\omega t} \ldots\ldots\ldots\ldots\ldots\ldots\ldots(27)$$

where the real parts of the quantities are retained. If

$$k = \xi + i\eta$$

where η is positive, the oscillations of the vector E are damped owing to the exponential factor $e^{-\eta x}$.

The elementary electromagnetic fields that have just been found are fundamental in the theory of the reflection and refraction of light at a plane surface. This theory is given in the text-books on Physical Optics* and need not be reproduced here. Various attempts have been made† to prove that any

* See for instance, Wood's *Physical Optics* (1911), Ch. 13; Jeans, *Electricity and Magnetism* (1911), Ch. 18.

† See for instance a series of papers by G. Johnstone Stoney, *Phil. Mag.* (5), Vol. 43 (1897), pp. 139, 273, 368; Vol. 44, pp. 98, 206; *Brit. Assn. Reports* (1902), p. 539; *Phil. Mag.* Feb. 1903. The idea is probably due to Stokes.

electromagnetic disturbance in the aether can be represented as the sum of a finite or infinite number of elementary disturbances of the character of plane-waves travelling in various directions. Such a representation is generally only suitable within a restricted domain of the variables x, y, z, t; nevertheless, it may sometimes be employed with advantage.

When waves of all directions and frequencies are considered, the method of summation leads to Whittaker's formula*

$$\Omega = \int_0^\pi \int_0^{2\pi} f[x \sin\alpha \cos\beta + y \sin\alpha \sin\beta + z \cos\alpha - ct, \alpha, \beta] \times \sin\alpha \, . \, d\alpha d\beta \quad \text{.........................(28)}$$

for a wave-function. The case when

$$f[\xi, \alpha, \beta] = e^{i\nu\xi} \quad \xi < \theta < \pi$$
$$= 0 \quad \xi > \theta$$

has been used by Debye† in a discussion of the behaviour of waves of light in the vicinity of a focus. In order that an integral of the type (28) may represent a wave-function it is not necessary for the limits of integration to be those chosen. The limits for α may, for instance, be 0 and θ where θ is a root of an equation of type

$$x \sin\theta \cos\beta + y \sin\theta \sin\beta + z \cos\theta - ct = F(\theta).$$

In order to obtain other types of elementary solutions it is necessary to transform our differential equations to a system of orthogonal coordinates (u, v, w) for which the linear element is given by

$$ds = \frac{du^2}{U^2} + \frac{dv^2}{V^2} + \frac{dw^2}{W^2} \quad \text{..................(29).}$$

If H_u, H_v, H_w are the three components of a vector H in directions normal to the surfaces $u =$ const., $v =$ const., $w =$ const. through a point (x, y, z), the corresponding components of rot H are of the type‡

$$VW\left[\frac{\partial}{\partial v}\left(\frac{H_w}{W}\right) - \frac{\partial}{\partial w}\left(\frac{H_v}{V}\right)\right] \quad \text{..............(30),}$$

* *Math. Ann.* (1903). See also G. N. Watson, *Mess. of Math.* Vol. 36 (1906), p. 98.

† *Ann. d. Phys.* Vol. 30 (1909), p. 735.

‡ See for instance, H. M. Macdonald, *Electric Waves*, Ch. 6; M. Abraham, *Math. Ann.* Bd. 52, p. 81. Some very general transformation-formulae are

the new expression for div H is

$$UVW\left[\frac{\partial}{\partial u}\left(\frac{H_u}{VW}\right)+\frac{\partial}{\partial v}\left(\frac{H_v}{WU}\right)+\frac{\partial}{\partial w}\left(\frac{H_w}{UV}\right)\right] \ldots (31),$$

and the wave-equation becomes*

$$UVW\left[\frac{\partial}{\partial u}\left(\frac{U}{VW}\frac{\partial\phi}{\partial u}\right)+\frac{\partial}{\partial v}\left(\frac{V}{WU}\frac{\partial\phi}{\partial v}\right)+\frac{\partial}{\partial w}\left(\frac{W}{UV}\frac{\partial\phi}{\partial w}\right)\right]$$
$$=\frac{1}{c^2}\frac{\partial^2\phi}{\partial t^2} \ldots\ldots\ldots (32).$$

It is also sometimes advantageous to transform the wave-equation to a system of coordinates for which

$$dx^2+dy^2+dz^2-c^2dt^2=A^2d\xi^2+B^2d\eta^2+C^2d\zeta^2-D^2d\tau^2,$$

the wave-equation then becomes

$$\frac{\partial}{\partial \xi}\left(\frac{BCD}{A}\frac{\partial\phi}{\partial \xi}\right)+\frac{\partial}{\partial \eta}\left(\frac{CDA}{B}\frac{\partial\phi}{\partial \eta}\right)+\frac{\partial}{\partial \zeta}\left(\frac{DAB}{C}\frac{\partial\phi}{\partial \zeta}\right)$$
$$=\frac{\partial}{\partial \tau}\left(\frac{ABC}{D}\frac{\partial\phi}{\partial \tau}\right) \ldots\ldots\ldots (33).$$

§ 12. The generalisation of wave-functions.

When a solution of the wave-equation has been obtained other solutions may be derived from it in various ways. For instance, the function obtained by differentiating the given wave-function any number of times with regard to the coordinates x, y, z, t, is also a wave-function. By adding together arbitrary constant multiples of all the wave-functions obtained in this way we may obtain a very general type of wave-function.

Another method of generalisation is to make an arbitrary change of rectangular axes. The wave-equation is a covariant for such a transformation and so is a wave-function of the new coordinates. A number of arbitrary constants can be introduced into the solution in this way. We can also make a linear transformation of coordinates for which the expression

$$dx^2+dy^2+dz^2-c^2dt^2$$

contained in the papers of V. Volterra, *Rend. Lincei*, Ser. 4, Vol. 5, pp. 599, 630 (1889), and J. Larmor, *Cambr. Phil. Trans.* Vol. 14 (1885), p. 121.

* Lamé, *Journ. de l'École Polytechnique*, Cah. 23 (1833), p. 215; *Leçons sur les coordonnées curvilignes*, t. 2. A simplified proof was published by Lord Kelvin, *Cambr. Math. Journ.* Vol. 4 (1843).

remains unaltered in form and the preceding remarks still hold good.

To illustrate this let us first of all add together two particular cases of Euler's wave-function $\frac{1}{r} f(r \pm ct)$, viz.

$$\frac{1}{r(r-ct)} \text{ and } \frac{1}{r(r+ct)},$$

we then see that $(r^2 - c^2t^2)^{-1}$ is a wave-function. Generalising this by writing $x - x_0$, $y - y_0$, $z - z_0$, $t - t_0$ in place of x, y, z, t, we obtain the wave-function

$$\frac{1}{(x-x_0)^2 + (y-y_0)^2 + (z-z_0)^2 - c^2(t-t_0)^2} \quad \text{......(34).}$$

When we have obtained a wave-function involving one or more arbitrary parameters we may obtain others from it by differentiating with regard to the parameters, or by integrating with regard to them after having multiplied the expression by an arbitrary function of the parameters. For instance, from the above wave-function we may derive the more general wave-function

$$\int \frac{f(\tau)\, d\tau}{(x-x_0)^2 + (y-y_0)^2 + (z-z_0)^2 - c^2(t-\tau)^2} \quad \text{......(35),}$$

where the integration is between constant limits. It is not, however, really necessary for the limits to be constant, we may for instance take them to be $-\infty$ and $t - \frac{1}{c}(r + r_0)$, where $r_0^2 = x_0^2 + y_0^2 + z_0^2$. The resulting integral is then a wave-function provided $f(\tau)$ behaves in a suitable manner. If we take $f(\tau) = 1$, we obtain a wave-function

$$u = \frac{1}{R} \log \frac{R - r - r_0}{R + r + r_0} \quad \text{.................(36),}$$

where $$R^2 = (x-x_0)^2 + (y-y_0)^2 + (z-z_0)^2.$$

This function is independent of t and so must be a solution of Laplace's equation $\Delta u = 0$. It is closely connected with the function used on p. 3 of Basset's *Hydrodynamics*, Vol. 2.

§ 13. **Transformations.**

In addition to the linear transformations that have already been mentioned there are certain other transformations which enable us to pass from one wave-function to another*.

The first transformation, which is analogous to inversion, is defined by the equations

$$x' = \frac{x}{s^2}, \quad y' = \frac{y}{s^2}, \quad z' = \frac{z}{s^2}, \quad t' = \frac{t}{s^2} \quad \text{...........(37)}$$

where $$s^2 = r^2 - c^2t^2.$$

It is easy to verify that if $f(x, y, z, t)$ is a wave-function, then

$$\frac{1}{s^2} f\left(\frac{x}{s^2}, \frac{y}{s^2}, \frac{z}{s^2}, \frac{t}{s^2}\right) \quad \text{..................(38)}$$

is also a wave-function†.

The second transformation is‡

$$x' = \frac{x}{z - ct}, \quad y' = \frac{y}{z - ct}, \quad z' = \frac{r^2 - c^2t^2 - 1}{2(z - ct)}, \quad t' = \frac{r^2 - c^2t^2 + 1}{2c(z - ct)}$$
.........(39).

It is easy to verify that if $f(x, y, z, t)$ is a wave-function, then

$$\frac{1}{z - ct} f\left[\frac{x}{z - ct}, \frac{y}{z - ct}, \frac{s^2 - 1}{2(z - ct)}, \frac{s^2 + 1}{2c(z - ct)}\right] \text{...(40)}$$

is also a wave-function. Since $e^{-(z+ct)}$ is a wave-function we may deduce in this way that

$$\frac{1}{z - ct} e^{-\frac{s^2}{z - ct}}$$

is a wave-function.

It should be noticed that if we put $4\tau = z - ct$, $\sigma = z + ct$, a function of the type

$$u = F(x, y, \tau) e^{-\sigma} \quad \text{.....................(41)}$$

* For a general account of these transformations see a paper by the author, *Proc. London Math. Soc.* Ser. 2, Vol. 7 (1908).

† This of course is a simple generalisation of Kelvin's theorem for Laplace's equation. The generalisation to the corresponding equation in n variables is mentioned by Bôcher, *Bull. of the Amer. Math. Soc.* Vol. 9 (1903), p. 459.

‡ *Proc. London Math. Soc.* Ser. 2, Vol. 7 (1909). This transformation is equivalent to a conformal transformation of a space of four dimensions which was discovered by Cremona. Cf. Darboux, *Leçons sur les systèmes orthogonaux et les coordonnées curvilignes*, Paris (1910).

is a wave-function if F satisfies the differential equation

$$\frac{\partial^2 F}{\partial x^2}+\frac{\partial^2 F}{\partial y^2}=\frac{\partial F}{\partial \tau} \qquad \text{.....................(42).}$$

The wave-function we have just obtained indicates that

$$\frac{1}{\tau}e^{-\frac{x^2+y^2}{4\tau}}=F \qquad \text{.....................(43)}$$

is a solution of the above equation. This solution is fundamental in the theory of the conduction of heat. It is evident that any solution of the equation of the conduction of heat in two dimensions can be used to construct a wave-function.

Our second transformation theorem for the wave-equation also tells us that if $F(x, y, \tau)$ is a solution of the equation (42) the function

$$\frac{1}{\tau}e^{-\frac{x^2+y^2}{4\tau}}F\left(\frac{x}{\tau},\ \frac{y}{\tau},\ -\frac{1}{\tau}\right) \qquad \text{..............(44)}$$

is also a solution. This result is due to J. Brill* and Appell†.

The corresponding theorems for the two-dimensional wave-equation

$$\frac{\partial^2 V}{\partial x^2}+\frac{\partial^2 V}{\partial y^2}=\frac{1}{c^2}\frac{\partial^2 V}{\partial t^2} \qquad \text{..................(45)}$$

are first that if $f(x, y, t)$ is a wave-function and

$$s^2=x^2+y^2-c^2t^2,$$

the function

$$\frac{1}{s}f\left(\frac{x}{s^2},\ \frac{y}{s^2},\ \frac{t}{s^2}\right) \qquad \text{.....................(46)}$$

is also a wave-function. This is equivalent to Lord Kelvin's theorem for Laplace's equation if we simply replace ict by z.

The second theorem is that if $f(x, y, t)$ is a wave-function, then

$$\frac{1}{\sqrt{y-ct}}f\left[\frac{x}{y-ct},\ \frac{s^2-1}{2(y-ct)},\ \frac{s^2+1}{2c(y-ct)}\right] \qquad \text{......(47)}$$

is also a wave-function. Writing $4\tau=y-ct$, $\sigma=y+ct$ as before we find that $u=e^{-\sigma}F(x, \tau)$ is a wave-function if

$$\frac{\partial^2 F}{\partial x^2}=\frac{\partial F}{\partial \tau} \qquad \text{........................(48).}$$

* *Messenger of Mathematics* (1891).

† *Liouville's Journal*, Ser. 4, t. 8 (1894).

The second theorem can now be used to show that

$$F = \tau^{-\frac{1}{2}} e^{-\frac{x^2}{4\tau}} \quad \text{.......................(49)}$$

is a solution of this equation and that if $F(x, \tau)$ is one solution the function

$$\tau^{-\frac{1}{2}} e^{-\frac{x^2}{4\tau}} F\left(\frac{x}{\tau}, -\frac{1}{\tau}\right) \quad \text{..................(50)}$$

is also a solution. This theorem is likewise due to Brill and Appell, it can evidently be generalised to the equation in n variables.

EXAMPLES.

1. Prove that it is possible for a train of plane electric waves to travel along an infinite isolated slab of dielectric material without being dissipated by spreading out into the adjacent empty space. Show that if $2a$ is the thickness and K the inductivity of the slab, the velocity of propagation of such waves of length λ along the slab, when polarised so that their magnetic vector is parallel to it, is

$$\frac{c}{\sqrt{K}}(1+\theta^2)^{\frac{1}{2}},$$

where θ is the lowest real or the pure imaginary root of the equation

$$\tan \frac{2\pi a}{\lambda}\theta = \left(\frac{K^2-K}{\theta^2} - K\right)^{\frac{1}{2}}.$$

(Larmor, Cambr. Math. Tripos, Part II, 1906.)

2. Prove that with the notation of § 13, the function

$$\frac{1}{(x-x_0)^2+(y-y_0)^2+(z-z_0)^2-c^2(t-t_0)^2} \frac{s+s_0}{\sqrt{ss_0+xx_0+yy_0+zz_0-c^2tt_0}}$$

is a wave-function, x_0, y_0, z_0, t_0 being arbitrary constants.

3. If $F(x, y, z, t)$ is a solution of the wave-equation, the function

$$V = \int_0^\infty e^{\frac{s^2}{4\tau} - \frac{\sigma^2\tau}{4} - \frac{\sigma ct}{2}} F\left(\frac{x}{\tau}, \frac{y}{\tau}, \frac{z}{\tau}, -\frac{t}{\tau}\right)\frac{d\tau}{\tau^2}$$

is, under suitable conditions, a solution of the equation

$$\Delta V = \frac{1}{c^2}\frac{\partial^2 V}{\partial t^2} + \frac{\sigma}{c}\frac{\partial V}{\partial t}.$$

The quantity s^2 has the meaning assigned to it in § 13 and is supposed in the present case to be negative.

4. The function

$$V=\frac{2}{\sqrt{\pi}}\int_{-\infty}^{\frac{1}{2}[\sqrt{v(x-vt+iy)}+\sqrt{v(x-vt-iy)}]} e^{-\lambda^2}\,d\lambda$$

satisfies the equation

$$\frac{\partial^2 V}{\partial x^2}+\frac{\partial^2 V}{\partial y^2}=\frac{\partial V}{\partial t},$$

of the conduction of heat in two dimensions: it is zero over a semi-infinite line which covers part of the axis of x and moves in the positive direction with uniform velocity v. The isothermal lines at any given instant are confocal parabolas.

5. Prove that, under suitable limitations, the function

$$V=\int_0^\infty e^{-\frac{\rho^2}{4t}-\lambda^2 t}\left[f\left(\frac{x+iy}{t}\right)+F\left(\frac{x-iy}{t}\right)\right]\frac{dt}{t}, \qquad \rho^2=x^2+y^2,$$

is a solution of

$$\frac{\partial^2 V}{\partial x^2}+\frac{\partial^2 V}{\partial y^2}=\lambda^2 V.$$

Obtain in this way the particular solution

$$\frac{1}{\rho}e^{-\lambda\rho}(x+iy)^{-\frac{1}{2}}.$$

6. Prove that if $x^2+y^2>t^2$, the integral

$$\int_0^\infty \frac{e^{-a\sigma}\,da}{x^2+y^2+4a^2-4at}=V$$

satisfies

$$\frac{\partial^2 V}{\partial x^2}+\frac{\partial^2 V}{\partial y^2}=\frac{\partial^2 V}{\partial t^2}+\sigma\frac{\partial V}{\partial t},$$

and if $x^2>t^2$, the integral

$$\int_0^\infty \frac{e^{-a\sigma}\,da}{\sqrt{x^2+4a^2-4at}}=V$$

satisfies

$$\frac{\partial^2 V}{\partial x^2}=\frac{\partial^2 V}{\partial t^2}+\sigma\frac{\partial V}{\partial t}.$$

7. Prove that if $\rho^2=x^2+y^2$, $\rho_0{}^2=x_0{}^2+y_0{}^2$, the integral

$$V=\int_{\rho+\rho_0}^\infty \frac{\sin k(\xi+a)\,d\xi}{\sqrt{\xi^2-(x-x_0)^2-(y-y_0)^2}}$$

satisfies the equation

$$\frac{\partial^2 V}{\partial x^2}+\frac{\partial^2 V}{\partial y^2}+k^2 V=0$$

CHAPTER III

POLAR COORDINATES

§ 14. The elementary solutions.

If we make Laplace's transformation

$$x = r \sin\theta \cos\phi, \quad y = r \sin\theta \sin\phi, \quad z = r\cos\theta \quad \ldots(51),$$

the equation $\Delta u + k^2 u = 0$ becomes

$$\frac{\partial^2 u}{\partial r^2} + \frac{2}{r}\frac{\partial u}{\partial r} + \frac{1}{r^2 \sin\theta}\frac{\partial}{\partial\theta}\left(\sin\theta \frac{\partial u}{\partial\theta}\right) + \frac{1}{r^2\sin^2\theta}\frac{\partial^2 u}{\partial\phi^2} + k^2 u = 0 \ldots(52).$$

This is satisfied by a function of the form

$$u = R(r)\,\Theta(\theta)\,\Phi(\phi)$$

if

$$\frac{\partial^2\Phi}{\partial\phi^2} + m^2\Phi = 0 \quad \ldots\ldots\ldots\ldots\ldots\ldots\ldots(53),$$

$$\frac{1}{\sin\theta}\frac{d}{d\theta}\left(\sin\theta\frac{d\Theta}{d\theta}\right) + \left[n(n+1) - \frac{m^2}{\sin^2\theta}\right]\Theta = 0 \ldots(54),$$

$$\frac{d^2R}{dr^2} + \frac{2}{r}\frac{dR}{dr} + \left[\frac{k^2}{r^2} - n(n+1)\right]R = 0 \quad \ldots\ldots(55).$$

The first equation is satisfied by $\Phi = \cos(m\phi + \alpha)$, the second by $P_n^m(\cos\theta)$ and $Q_n^m(\cos\theta)$, where these are the associated Legendre functions. The third equation may be written in Bessel's form

$$\frac{d^2w}{dr^2} + \frac{1}{r}\frac{dw}{dr} + \left[\frac{k^2}{r^2} - (n+\tfrac{1}{2})^2\right]w = 0 \quad \ldots\ldots(56),$$

where $w = r^{\frac{1}{2}}R$, and is satisfied by $J_{n+\frac{1}{2}}(kr)$ and $J_{-(n+\frac{1}{2})}(kr)$.

In these solutions m and n can have any constant values. It should be noticed that when $n+\frac{1}{2}$ is an integer the Bessel functions that have just been written down are not independent and the second solution $Y_{n+\frac{1}{2}}(kr)$ of Bessel's equation must be used.

In dealing with a problem such as the effect of an obstacle on a train of electric waves, the secondary waves sent out from the obstacle must have the character of diverging waves at a great distance from the obstacle. In the case $n = 0$ the differential equation for R is satisfied by

$$R = \frac{1}{r} e^{\pm ikr} \qquad \text{.....................(57)},$$

and if the real part of k has the same sign as ω when the electric and magnetic forces are the real parts of expressions of the form $Ae^{-i\omega t}$, we can obtain a solution appropriate for the representation of a diverging wave by taking the positive sign, for then we have a function of the form

$$\frac{1}{r} e^{i(kr - \omega t)}.$$

Neither of the given solutions of Bessel's equation has the required form in fact

$$J_{\frac{1}{2}}(kr) = \sqrt{\frac{2}{\pi kr}} \sin kr, \quad J_{-\frac{1}{2}}(kr) = \sqrt{\frac{2}{\pi kr}} \cos kr.$$

We may, however, obtain solutions of the form (57) by taking a suitable combination of the preceding solutions.

In the case of electromagnetic fields in the free aether the physical interpretation of the elementary wave-functions when n is zero is as follows* :

$\frac{1}{r} \cos k(r - ct)$	Progressive divergent waves.
$\frac{1}{r} \cos k(r + ct)$	Progressive convergent waves.
$\frac{1}{r} \cos kr \, . \cos kct$	Standing forced waves, source at origin.
$\frac{1}{r} \sin kr \, . \cos kct$	Standing free waves.

To obtain the solution of (56) appropriate for divergent waves when n has any value we write†

* A fuller discussion is given by A. Sommerfeld, *Jahresbericht der deutsch. math. Verein*, Bd. 21 (1913).

† The theory is due to Stokes, *Phil. Trans.* Vol. 158 (1868), p. 447; *Collected Papers*, Vol. 4, p. 321. See also Rayleigh's *Sound*, Vol. 2, p. 304. It should be mentioned that different notations are used by different writers. This is the notation used by Debye, *Ann. d. Phys.* Vol. 30 (1909), p. 57.

$$\left.\begin{aligned}\psi_n(x)&=\sqrt{\frac{\pi x}{2}}\,J_{n+\frac{1}{2}}(x)\\ \chi_n(x)&=(-)^n\sqrt{\frac{\pi x}{2}}\,J_{-n-\frac{1}{2}}(x)=-Y_{n+\frac{1}{2}}(x)\sqrt{\frac{\pi x}{2}}\\ \eta_n(x)&=\psi_n(x)-i\chi_n(x)\\ \zeta_n(x)&=\psi_n(x)+i\chi_n(x)\end{aligned}\right\}\ldots(58).$$

These new functions η_n, ζ_n are connected with Hankel's cylindrical functions* by the relations

$$\eta_n(x)=\sqrt{\frac{\pi x}{2}}\,H_1^{n+\frac{1}{2}}(x),\quad \zeta_n(x)=\sqrt{\frac{\pi x}{2}}\,H_2^{n+\frac{1}{2}}(x).$$

When the real part of x is large and positive we have the asymptotic expansion

$$\eta_n(x)=(-i)^{n+1}e^{ix}\left[1+\frac{i}{2x}\frac{n(n+1)}{1!}\right.$$
$$\left.-\frac{1}{(2x)^2}\frac{(n-1)n(n+1)(n+2)}{2!}+\ldots\right]\ldots(59).$$

The series terminates when n is an integer and then gives a true representation of the function. To get $\zeta_n(x)$ we change the sign of i. Various other notations have been used for the solution of Bessel's equation that is suitable for the representation of diverging waves. Lamb uses $D_\nu(x)$, $(\nu=n+\frac{1}{2})$ to denote the solution of equation (56) which has the asymptotic value

$$D_\nu(x)=\left(\frac{2}{\pi x}\right)^{\frac{1}{2}}i^\nu e^{-i\left(x+\frac{\pi}{4}\right)},$$

while many other English writers use $K_\nu(ix)$ to denote the solution with the asymptotic form

$$\left(\frac{\pi}{2x}\right)^{\frac{1}{2}}e^{i\left(\frac{\pi}{4}-x\right)}.$$

The following formulae will be found useful:

$$\psi_n(kr)=\frac{(kr)^{n+1}}{1.3\ldots(2n+1)}$$
$$\times\left[1-\frac{(kr)^2}{2(2n+3)}+\frac{(kr)^4}{2.4.(2n+3)(2n+5)}-\ldots\right]\ldots\ldots(60),$$

* See Nielsen's *Handbuch der Cylinderfunktionen*, p. 16.

$$\zeta_n(kr) = i^{n+1}e^{-ikr}\left[1 - \frac{i}{2kr}\frac{n(n+1)}{1}\right.$$
$$\left.- \frac{1}{(2kr)^2}\cdot\frac{(n-1)n(n+1)(n+2)}{2!} + \ldots (-i)^n\cdot\frac{(2n)!}{(2kr)^n . n!}\right] \ldots(61).$$

We have written down the last term in the series on the supposition that n is an integer. In this case

$$\psi_n(kr) = \tfrac{1}{2}[i^{n+1}e^{-ikr} + (-i)^{n+1}e^{ikr}], \quad |r| \text{ large} \ldots(62),$$

$$(2n+1)\frac{d\psi_n(x)}{dx} = (n+1)\psi_{n-1}(x) - n\psi_{n+1}(x) \quad \ldots(63),$$

$$(2n+1)\psi_n(x) = x[\psi_{n-1}(x) + \psi_{n+1}(x)] \quad \ldots\ldots\ldots(64),$$

$$K_\nu(x) = \int_0^\infty e^{-x\cosh\alpha}\cosh\nu\alpha . d\alpha \quad \ldots\ldots\ldots\ldots(65),$$

$$J_\nu(x) = \frac{x^\nu}{2^\nu\Gamma(\frac{1}{2})\Gamma(\nu+\frac{1}{2})}\int_0^\pi \cos(x\cos\alpha)\sin^{2\nu}\alpha . d\alpha, \quad R(\nu) > -\tfrac{1}{2} \quad \ldots\ldots\ldots\ldots(66),$$

$$\left.\begin{aligned}\int_{-\infty}^\infty J_{n+\frac{1}{2}}(kr)J_{p+\frac{1}{2}}(kr)\frac{dr}{r} &= 0 \qquad (n \neq p)\\ &= \frac{1}{2n+1}\ (n = p)\end{aligned}\right\} n \geqslant 0, p > 0 \quad (67).$$

In the last formula n and p are supposed to be integers. For further properties of Bessel functions the reader should consult Gray and Mathews' *Treatise on Bessel Functions*, Nielsen's *Handbuch der Cylinderfunktionen*, and Whittaker's *Analysis*. Tables are given in the first work and in Jahnke and Emde's *Funktionentafeln*, Leipzig (Teubner). A few additions to the tables have been made recently by J. W. Nicholson, *Proc. London Math. Soc.* Ser. 2, Vol. 11, p. 104; Dinnik, *Archiv der Mathematik und Physik* (3), Bd. 20, Heft 3, 1912; J. R. Airey, *Phil. Mag.* Vol. 22 (1911), p. 85, *Brit. Ass. Reports* (1911); A. Lodge, *Brit. Ass. Reports* (1909); J. G. Isherwood, *Manchester Memoirs* (1904).

The best definitions of the generalised Legendre functions for unrestricted values of m and n are those given by Hobson*.

* *Phil. Trans.* A, Vol. 187 (1896), pp. 443—531. E. W. Barnes has recently given new definitions of the functions as integrals involving Gamma Functions which make it possible for the principal formulae to be proved very quickly. His definition of $Q_n^m(x)$ differs from that of Hobson by a numerical factor which becomes rather troublesome when n is an integer and m is not.

We are interested here in the case when the variable is $\cos\theta$ and θ is a real angle, such that $0 < \theta < \pi$. We may then put, with the usual notation of the Gamma and hypergeometric functions,

$$P_n^m(\cos\theta) = \frac{1}{\Gamma(1-m)} \cot^m \frac{\theta}{2}\, F\left(-n, n+1;\ 1-m;\ \sin^2\frac{\theta}{2}\right) \quad \text{......(68)},$$

$$Q_n^m(\cos\theta) = \frac{\pi}{2\sin(n+m)\pi} \frac{1}{\Gamma(1-m)} \times \left\{\cos(n+m)\pi \cot^m \frac{\theta}{2}\, F\left(-n, n+1;\ 1-m;\ \sin^2\frac{\theta}{2}\right) - \tan^m \frac{\theta}{2}\, F\left(-n, n+1, 1-m, \cos^2\frac{\theta}{2}\right)\right\} \quad \text{.........(69)},$$

$$P_n^{-m}(\cos\theta) = \frac{\operatorname{cosec}^m\theta}{2^m\Gamma(\tfrac{1}{2})\Gamma(m+\tfrac{1}{2})} \int_0^\theta \frac{\cos(n+\tfrac{1}{2})\phi}{(2\cos\phi - 2\cos\theta)^{\frac{1}{2}-m}}, \quad R(m+\tfrac{1}{2}) > 0 \quad \text{......(70)}.$$

When m is a positive integer, we have

$$\left.\begin{aligned} P_n^m(x) &= (1-x^2)^{\frac{m}{2}} \frac{d^m}{dx^m} P_n(x) \\ Q_n^m(x) &= (1-x^2)^{\frac{m}{2}} \frac{d^m}{dx^m} Q_n(x) \end{aligned}\right\} \quad \text{......(71)},$$

and when n is a positive integer

$$P_n(x) = \frac{1.3\ldots(2n-1)}{1.2\ldots n}\left[x^n - \frac{n(n-1)}{2(2n-1)} x^{n-2} + \frac{n(n-1)(n-2)(n-3)}{2.4(2n-1)(2n-3)} x^{n-4} - \ldots\right] \quad \text{......(72)},$$

$$P_n(\cos\theta) = \cos^n\theta - \frac{n(n-1)}{2^2}\cos^{n-2}\theta . \sin^2\theta + \frac{n(n-1)(n-2)(n-3)}{2^2 . 4^2}\cos^{n-4}\theta . \sin^4\theta \ldots \quad \text{......(73)},$$

$$\frac{1}{r^{n+1}} P_n(\cos\theta) = \frac{(-1)^n}{n!} \frac{d^n}{dz^n}\left(\frac{1}{r}\right) \quad \text{..................(74)},$$

$$\frac{1}{r^{n+1}} Q_n(\cos\theta) = \frac{(-1)^n}{n!} \frac{d^n}{dz^n}\left(\frac{1}{2r}\log\frac{r+z}{r-z}\right) \quad \text{......(75)}.$$

The last two formulae illustrate the method of deriving more complicated wave-functions from simple ones by differentiation. The functions

$$\frac{1}{r} \text{ and } \frac{1}{2r} \log \frac{r+z}{r-z}$$

are in fact solutions of (52) when $k = 0$.

The formula

$$\int_{-1}^{+1} P_n^m(x) P_\nu^m(x) dx = 0 \quad (n \neq \nu)$$
$$= \frac{2}{2n+1} \cdot \frac{(n+m)!}{(n-m)!} \quad \ldots\ldots(76),$$

in which m, n, ν are positive integers or zero, enables the coefficients in an expansion of a function in series of functions $P_n^m(x)$ to be determined by a simple integration.

For further properties of the Legendre functions we must refer to Heine's *Kugelfunktionen*, Byerly's *Fourier Series and Spherical Harmonics*, Whittaker's *Analysis*, Nielsen's *Handbuch der Cylinderfunktionen* and *Théorie des fonctions métasphériques*, and to memoirs by E. W. Hobson* and E. W. Barnes†.

Tables of the Legendre functions have been published by J. W. L. Glaisher‡, J. Perry§ and A. Lodge‖; some tables of the functions $P_n^m(\mu)$ have been given by H. Tallquist¶.

For the history of the functions of Legendre and Bessel the reader should consult the article by A. Wangerin in the *Encyklopädie der Mathematischen Wissenschaften*, Bd. II. 1, Heft 5 (1904), p. 695.

§ 15. Relations between various solutions.

We have already remarked that when a wave-function or a solution of equation (52) involving arbitrary constants has been found, other solutions may be derived from it by the method of summation or integration. By choosing our sum or integral so that it represents certain simple solutions of the

* *Proc. London Math. Soc.* Ser. 1, Vol. 22 (1891), p. 431, and *op. cit.*

† *Quarterly Journal*, Vol. 39 (1908), p. 97.

‡ *Brit. Ass. Report* (1879).

§ *Phil. Mag.*, Dec. (1891). See also Byerly, *loc. cit.*

‖ *Phil. Trans.* A, Vol. 203 (1904).

¶ *Acta Societatis Fennicae*, Vols. 32, 33.

fundamental equation, a number of important identities may be obtained. A few formulae will be written down to illustrate this*.

If $$R^2 = r^2 + r_1^2 - 2rr_1 \cos\theta,$$

$$\frac{\sin(kR)}{R} = \frac{\pi}{\sqrt{(rr_1)}} \sum_{n=0}^{\infty} (2n+1) J_{n+\frac{1}{2}}(kr) J_{n+\frac{1}{2}}(kr_1) P_n(\cos\theta)$$
(Heine and Hobson)......(77),

$$\frac{1}{R} e^{-ikR} = \frac{1}{rr_1} \sum_{n=0}^{\infty} (2n+1) \psi_n(kr) \zeta_n(kr_1) P_n(\cos\theta) \quad (r_1 > r)$$
(Heine and Macdonald)......(78),

$$r^{-\frac{1}{2}} J_{n+\frac{1}{2}}(kr) P_n(\cos\theta) = \frac{1}{\pi} \int_{-\infty}^{\infty} \frac{\sin(kR)}{R} J_{n+\frac{1}{2}}(kr_1) r_1^{-\frac{1}{2}} dr_1$$
.........(79).

We may illustrate the peculiar behaviour of certain definite integral solutions of our fundamental equation by the following example, in which k is supposed to be real and positive.

Let $$f_m(x) = \int_0^m \cos xt\, \chi(t)\, dt,$$

where $\chi(t)$ is a function such that $\int_0^m |\chi(t)|\, dt$ is convergent, then it may be proved by means of Fourier's double integral theorem that†

$$\frac{1}{\pi} \int_{-\infty}^{\infty} \frac{\sin k(r - r_1)}{r - r_1} f_m(r_1)\, dr_1 = f_m(r) \quad (k > m)$$
$$= f_k(r) \quad (k \leqslant m).$$

Hence $$u = \frac{1}{\pi} \int_{-\infty}^{\infty} \frac{\sin kR}{R} f_m(r_1)\, dr_1$$

is a solution of (52) which reduces to either $f_m(r)$ or $f_k(r)$ when $\theta = 0$. Solutions of (52) which are derived from the elementary solutions by integrating with regard to n have been employed by H. W. March‡. He makes use of an inversion-formula§

* Some very general formulae are given by L. Gegenbauer, *Monatsh. f. Math.* Bd. 10 (1899), p. 189.

† This equation is obtained in a different manner by G. H. Hardy, *Proc. London Math. Soc.* (2), Vol. 7 (1909), p. 445.

‡ *Ann. d. Physik*, Bd. 37 (1912). See also H. Poincaré, *Comptes Rendus*, t. 154 (1912), p. 795; W. v. Rybczynski, *Ann. d. Phys.* Bd. 41 (1913).

§ This is in some respects analogous to the inversion-formula given by F. G. Mehler, *Math. Ann.* Bd. 18 (1881), p. 161.

$$\left.\begin{aligned} f(\theta) &= \int_0^\infty P_{\alpha-\frac{1}{2}}(\cos\theta)\,\psi(\alpha)\,\alpha d\alpha \\ \psi(\alpha) &= \int_0^\pi S_{\alpha-\frac{1}{2}}(\cos\gamma) f(\gamma) \sin\gamma d\gamma \end{aligned}\right\} \quad \text{......(80)},$$

where

$$S_{\alpha-\frac{1}{2}}(\cos\gamma) = \frac{\cos\alpha\pi}{\alpha\pi\cos\frac{\gamma}{2}} + \frac{2}{\pi}\int_\gamma^\pi \frac{\sin\alpha\beta\,.\,d\beta}{\{2(\cos\gamma-\cos\beta)\}^{\frac{1}{2}}} \text{......(81)}.$$

It is sometimes instructive to find how a wave-function, depending on an arbitrary function, can be expressed in terms of elementary wave-functions. Now in the second example of § 5 the electric and magnetic forces are all of the form

$$\frac{1}{r} f\left[r - ct, \frac{x \pm iy}{z+r}\right] \quad \text{..............(82)},$$

or are the sums of terms of this form. Consequently, a function of this type may be expected to be a wave-function and it is easy to verify that this is the case*. We may now deduce that

$$\frac{1}{r} e^{\pm ikr} F\left(\frac{x \pm iy}{z+r}\right) \text{....................(83)}$$

is a solution of (52) and consequently it follows that $\tan^m \frac{\theta}{2}$ and $\cot^m \frac{\theta}{2}$ are solutions of equation (54) when $n = 0$. We have in fact

$$P_0^m(\cos\theta) = \frac{1}{\Gamma(1-m)} \cot^m \frac{\theta}{2},$$

$$Q_0^m(\cos\theta) = \tfrac{1}{2}\Gamma(m)\left(\cos\overline{m\pi}\,.\cot^m\frac{\theta}{2} - \tan^m\frac{\theta}{2}\right).$$

In the last formula m must not be zero or a negative integer.

§ 16. **The convergence of series of elementary solutions.**

When $k = 0$ our fundamental equation (52) reduces to Laplace's equation $\Delta u = 0$ and we have the familiar elementary solutions

* The theorem also follows immediately from a result given by A. R. Forsyth, *Messenger of Mathematics* (1898), p. 114, and E. W. Hobson, *loc. cit.*

$$\left.\begin{array}{ll} r^n P_n^m(\cos\theta)\cos m(\phi-\phi_0), & r^n Q_n^m(\cos\theta)\cos m(\phi-\phi_0) \\ \dfrac{1}{r^{n+1}} P_n^m(\cos\theta)\cos m(\phi-\phi_0), & \dfrac{1}{r^{n+1}} Q_n^m(\cos\theta)\cos m(\phi-\phi_0) \end{array}\right\} \quad \text{.........(84).}$$

A pair of series of the type

$$\Sigma\left(\frac{r}{a}\right)^n f_n(\theta, \phi), \quad \Sigma\left(\frac{a}{r}\right)^{n+1} F_n(\theta, \phi)$$

which converge when $r=a$ are suitable for representing harmonic functions inside and outside the sphere $r=a$ because the first converges absolutely when $r<a$ and the second when $r>a$.

The case in which $k \neq 0$ is very similar. When n is large the function $\psi_n(kr)$ may be replaced by

$$\frac{(kr)^{n+1}}{1.3\ldots(2n+1)}$$

and so a series of the form

$$\sum_{n=0}^{\infty} \psi_n(kr) f_n(\theta, \phi)$$

converges like a power series. Again, when kr is real, we have

$$\begin{aligned} |\zeta_n(kr)|^2 = 1 &+ \frac{1}{2} n(n+1)\frac{1}{(kr)^2} \\ &+ \frac{1.3}{2.4}(n-1)n(n+1)(n+2) . \frac{1}{(kr)^4} + \ldots \\ &+ \frac{1.3\ldots(2n-1)}{2.4\ldots(2n)}(2n!)\frac{1}{(kr)^{2n}} \quad \text{.........(85),} \end{aligned}$$

n being a positive integer. It is clear from this equation that $|\zeta_n(kr)|$ decreases as r increases, hence if a series of the form

$$\sum_{n=0}^{\infty} \zeta_n(kr) f_n(\theta, \phi)$$

converges absolutely for any value of r it converges absolutely for all greater values of r.

For a discussion of the convergence of series of spherical harmonics we may refer to C. Neumann's book *Ueber die nach Kreis-, Kugel- und Cylinder-Funktionen fortschreitenden Entwickelungen*, Leipzig (1881); to Heine's *Kugelfunktionen*, Bd. 1, p. 435, Bd. 2, p. 361, and to papers by U. Dini, *Ann. di Mat.* (2), t. 6 (1874); H. Poincaré, *Comptes Rendus*, t. 118 (1894), p. 497; S. Chapman, *Quarterly Journal*, Vol. 43 (1912), p. 1; T. H.

Gronwall, *Math. Ann.* Vol. 74 (1913), Vol. 75 (1914), *Comptes Rendus* (1914), *Amer. Trans.* Jan. (1914), Vol. 15; C. Jordan, *Cours d'Analyse*, 2nd ed., Vol. 2, p. 252; and B. H. Camp, *Bull. of the Amer. Math. Soc.* Vol. 18 (1912), p. 236. The convergence of series of Legendre polynomials has been discussed very thoroughly by G. Darboux, *Liouville's Journal* (2), t. 19 (1874), p. 1; (3), t. 4, p. 393; O. Blumenthal, *Dissertation*, Göttingen (1898); E. W. Hobson, *Proc. London Math. Soc.* (2), Vol. 7 (1909); L. Féjer, *Math. Ann.* Bd. 67, p. 76; D. Jackson, *Amer. Trans.* Vol. 13 (1912).

§ 17. **The scattering of plane homogeneous electromagnetic waves by a spherical obstacle.**

The effect of small particles in scattering incident radiation has been discussed very thoroughly by Lord Rayleigh* who has used it as the basis of a mathematical theory of the blue colour of the sky. The action of a single spherical particle is of fundamental importance and so the electromagnetic theory of the scattering of light by a dielectric sphere has been worked out by Lord Rayleigh†, Prof. Love‡ and other writers. This theory can also be developed so as to cover the mathematical theory of the rainbow.

The more general theory of the scattering of incident radiation by a spherical obstacle§ with arbitrary optical properties|| admits of some very interesting applications in the study of the colours exhibited by metal glasses, metallic films and colloidal solutions or suspensions of metals. The electromagnetic theory of these colours has been developed by J. Maxwell Garnett¶, G. Mie**, R. Gans†† and Happel††, who have considered

* *Phil. Mag.* Vol. 41 (1871), pp. 107, 274, 447; Vol. 12 (1881), p. 81; *Collected Papers*, Vol. 1, pp. 87, 104, 518.

† *Phil. Mag.* Vol. 44 (1897), pp. 28—52; *Collected Papers*, Vol. 4, p. 321; *Proc. Roy. Soc.* Vol. 84 (1910), p. 25; Vol. 90 (1914), p. 219.

‡ *Proc. London Math. Soc.* Vol. 30 (1899), p. 308.

§ The work of Stokes, *Camb. Trans.* Vol. 9 (1849), p. 1, with later applications, *Collected Papers*, Vol. 4, and of L. Lorenz, *Wied. Ann.* Vol. 2 (1880), p. 70, opened up the subject.

|| The case of small conductivity was discussed by G. W. Walker, *Quart. Journ.* Vol. 30 (1899), p. 204; Vol. 31 (1900), p. 36.

¶ *Phil. Trans.* A, Vol. 203 (1904), p. 385; Vol. 205 (1905), p. 237.

** *Ann. d. Phys.* Vol. 25 (1908), p. 377.

†† *Ibid.* Vol. 29 (1909), p. 280; Vol. 37 (1912), p. 881.

the cases of spheres and ellipsoids endowed with the optical constants ϵ, μ, σ.

The particular case of a perfectly conducting sphere was worked out by J. J. Thomson* and has been discussed in greater detail by J. W. Nicholson†.

The problem is also of importance in connection with the theory of comets' tails which has been developed by Euler, FitzGerald‡ and Arrhenius§. The pressure of light on a perfectly conducting spherical obstacle has accordingly been calculated by K. Schwarzschild‖ and J. W. Nicholson¶. The more general case of a sphere with the optical constants ϵ, μ, σ has been treated very fully by P. Debye**.

Let us assume that the electric and magnetic forces E', H' at any point of space are the real parts of vectors E, H of the form $Ae^{i\omega t}$, where A is a complex quantity independent of t. We then write

$$H \pm i\nu E = Me^{i\omega t}, \text{ where } \nu^2 = \frac{\epsilon\omega - i\sigma}{\mu\omega} \qquad \text{.........(86)}$$

and we find that the differential equations satisfied by M in a medium whose optical constants are ϵ, μ, σ can be written in the form

$$\left.\begin{aligned} \pm kM_r &= \frac{1}{r^2 \sin\theta}\left[\frac{\partial}{\partial\theta}(r\sin\theta \,.\, M_\phi) - \frac{\partial}{\partial\phi}(rM_\theta)\right] \\ \pm kM_\theta &= \frac{1}{r\sin\theta}\left[\frac{\partial M_r}{\partial\phi} - \frac{\partial}{\partial r}(r\sin\theta \,.\, M_\phi)\right] \\ \pm kM_\phi &= \frac{1}{r}\left[\frac{\partial}{\partial r}(rM_\theta) - \frac{\partial M_r}{\partial\theta}\right] \end{aligned}\right\} \qquad \text{......(87)},$$

where $k = \dfrac{\mu\nu\omega}{c}$ or $k^2 = \dfrac{\epsilon\mu\omega^2 - i\mu\sigma\omega}{c^2}$.

* *Recent Researches*, p. 437.

† *Proc. London Math. Soc.* (2), Vol. 9 (1910), p. 67; Vol. 11 (1912), p. 277.

‡ *Scientific Writings*, pp. 108, 531.

§ *Phys. Zeitschr.* Vol. 2 (1901), pp. 81—97; *Das Werden der Welten*, Leipzig (1907), p. 85.

‖ *Sitzungsber. d. Kgl. Bayer. Akad. d. Wiss.* Vol. 31 (1901), p. 293.

¶ *British Association Reports* (1910), p. 544; *Monthly Notices of the Royal Astronomical Society*, Vol. 70, p. 544. See also J. Proudman, *Ibid.* Vol. 73 (1913), p. 535.

** *Ann. d. Physik*, Vol. 30 (1909), p. 57.

These equations may be satisfied by putting

$$\left.\begin{aligned} M_r &= \frac{\partial^2}{\partial r^2}(r\Omega) + k^2 r\Omega \\ M_\theta &= \frac{1}{r}\frac{\partial^2 (r\Omega)}{\partial r \partial\theta} \pm \frac{k}{r\sin\theta}\frac{\partial (r\Omega)}{\partial\phi} \\ M_\phi &= \frac{1}{r\sin\theta}\frac{\partial^2 (r\Omega)}{\partial r\partial\phi} \mp \frac{k}{r}\frac{\partial (r\Omega)}{\partial\theta} \end{aligned}\right\} \qquad (88),$$

where the function $\Omega = U \pm iV$ is a solution of (52).

The electric and magnetic forces may be derived at once from these expressions by equating the ambiguous and unambiguous parts as explained in § 2.

We shall now assume that the incident wave of plane homogeneous monochromatic polarised light is represented by

$$(M_r, M_\theta, M_\phi) = e^{ikr\cos\theta \pm i\phi}(\sin\theta, \cos\theta, \pm i) \qquad (89),$$

the electric vector being parallel to the axis of y.

The corresponding function Ω_0 is obtained by solving the equation

$$\frac{\partial^2}{\partial r^2}(r\Omega_0) + k^2 r\Omega_0 = \sin\theta \,.\, e^{ikr\cos\theta \pm i\phi}.$$

We easily find that

$$r\Omega_0 = k^{-2} e^{\pm i\phi}[\operatorname{cosec}\theta \,.\, e^{ikr\cos\theta} + f_1(\theta, \phi) e^{ikr} + f_2(\theta, \phi) e^{-ikr}] \qquad (90).$$

Choosing the unknown functions so that $r\Omega_0$ is finite for $\theta = 0$ and $\theta = \pi$ we obtain finally

$$r\Omega_0 = \tfrac{1}{2}k^{-2} e^{\pm i\phi}\left[2\operatorname{cosec}\theta \,.\, e^{ikr\cos\theta} - \cot\frac{\theta}{2}\, e^{ikr} - \tan\frac{\theta}{2}\, e^{-ikr}\right] \qquad (91).$$

We may now assume an expansion for $r\Omega_0$ of the form

$$r\Omega_0 = k^{-2} e^{\pm i\phi}\sum_{n=1}^{\infty} a_n \psi_n(kr) P_n^1(\cos\theta).$$

To determine the coefficients a_n we multiply by $\sin\theta$ and differentiate both sides of the equation. Then since

$$\frac{d}{d\theta}[\sin\theta \,.\, P_n^1(\cos\theta)] = -n(n+1)\sin\theta \,.\, P_n(\cos\theta) \qquad (92),$$

the coefficients may be determined at once with the aid of Lord Rayleigh's expansion*

$$ikr \sin\theta \, . \, e^{ikr\cos\theta} = \sum_{n=0}^{\infty} i^{n+1}(2n+1)\,\psi_n(kr)\,P_n(\cos\theta)\sin\theta \quad \text{.........(93).}$$

We thus obtain

$$r\Omega_0 = k^{-2} e^{\pm i\phi} \sum_{n=1}^{\infty} i^{n-1} \frac{2n+1}{n(n+1)}\,\psi_n(kr)\,P_n^1(\cos\theta) \quad \text{......(94).}$$

Now let Ω_1, Ω_2 be the functions from which the electric and magnetic forces in the scattered light and transmitted light may be derived respectively. The appropriate forms are given by equations of the type

$$\left. \begin{aligned} rU_1 &= \sum_{n=1}^{\infty} A_n \zeta_n(kr)\,P_n^1(\cos\theta)\cos\phi \\ rV_1 &= \sum_{n=1}^{\infty} B_n \zeta_n(kr)\,P_n^1(\cos\theta)\sin\phi \end{aligned} \right\} \quad \text{......(95),}$$

$$\left. \begin{aligned} rU_2 &= \sum_{n=1}^{\infty} C_n \psi_n(hr)\,P_n^1(\cos\theta)\cos\phi \\ rV_2 &= \sum_{n=1}^{\infty} D_n \psi_n(hr)\,P_n^1(\cos\theta)\sin\phi \end{aligned} \right\} \quad \text{......(96),}$$

where h, η are the values of k, ν respectively within the sphere.

It is easy to deduce from (88) that the tangential components of the electric and magnetic forces are continuous in crossing the sphere $r = a$, if when $r = a$

$$\left. \begin{aligned} \frac{k}{\nu}[U_0 + U_1] &= \frac{h}{\eta} U_2 \\ \frac{\partial}{\partial r}[r(U_0 + U_1)] &= \frac{\partial}{\partial r}(rU_2) \\ k(V_0 + V_1) &= hV_2 \\ \frac{1}{\nu}\frac{\partial}{\partial r}[r(V_0 + V_1)] &= \frac{1}{\eta}\frac{\partial}{\partial r}(rV_2) \end{aligned} \right\} \quad \text{.........(97).}$$

* *Theory of Sound*, Vol. 2, p. 272. The expansion was also obtained independently by Heine, *Kugelfunktionen* (1878), Bd. 1, p. 82.

These conditions give

$$\left.\begin{aligned}
\frac{k}{\nu} A_n \zeta_n(ka) - \frac{h}{\eta} C_n \psi_n(ha) &= \frac{1}{k\nu} i^{n+1} \frac{2n+1}{n(n+1)} \psi_n(ka) \\
k A_n \zeta_n'(ka) - h C_n \psi_n'(ha) &= \frac{1}{k} i^{n+1} \frac{2n+1}{n(n+1)} \psi_n'(ka) \\
k B_n \zeta_n(ka) - h D_n \psi_n(ha) &= \frac{1}{k} i^{n+1} \frac{2n+1}{n(n+1)} \psi_n(ka) \\
\frac{k}{\nu} B_n \zeta_n'(ka) - \frac{h}{\eta} D_n \psi_n'(ha) &= \frac{1}{k\nu} i^{n+1} \frac{2n+1}{n(n+1)} \psi_n'(ka)
\end{aligned}\right\} \quad \ldots(98).$$

Solving these we get

$$A_n = i^{n+1} k^{-2} \frac{2n+1}{n(n+1)} \frac{H_n}{G_n}, \quad C_n = \frac{\eta}{hk} i^{n+1} \frac{2n+1}{n(n+1)} \frac{F_n}{G_n},$$

$$B_n = i^{n+1} k^{-2} \frac{2n+1}{n(n+1)} \frac{E_n}{L_n}, \quad D_n = \frac{\eta}{hk} i^{n+1} \frac{2n+1}{n(n+1)} \frac{F_n}{L_n},$$

where

$$\left.\begin{aligned}
H_n &= \eta \psi_n(ka) \psi_n'(ha) - \nu \psi_n'(ka) \psi_n(ha) \\
E_n &= \nu \psi_n(ka) \psi_n'(ha) - \eta \psi_n'(ka) \psi_n(ha) \\
F_n &= \zeta_n'(ka) \psi_n(ka) - \zeta_n(ka) \psi_n'(ka) \\
G_n &= \eta \psi_n'(ha) \zeta_n(ka) - \nu \zeta_n'(ka) \psi_n(ha) \\
L_n &= \nu \psi_n'(ha) \zeta_n(ka) - \eta \zeta_n'(ka) \psi_n(ha)
\end{aligned}\right\} \quad \ldots(99).$$

We can prove that our series all converge absolutely and uniformly at about the same rate as a power series of the form

$$\Sigma \frac{x^n}{1.3\ldots(2n+1)},$$

where x is either kr or $\frac{ka^2}{r}$. The proof depends on the fact that when n is large we have approximately

$$\psi_n(x) \sim \frac{x^{n+1}}{1.3\ldots(2n+1)}, \quad \psi_n'(x) = \frac{(n+1)x^n}{1.3\ldots(2n+1)},$$

$$\zeta_n(x) \sim 1.3\ldots(2n-1) \frac{i}{x^n} e^{-ix},$$

$$\zeta_n'(x) = -1.3\ldots(2n-1) n i x^{-n-1} e^{-ix}.$$

§ 18. **Free damped vibrations for the space outside the sphere.**

It should be noticed that some of the terms of our series become infinite when either $G_n = 0$ or $L_n = 0$: fortunately, however, the roots of these equations turn out to be complex and so when k is real no values of k need be excluded from the discussion. The damped vibrations determined by the roots of the equations $L_n = 0$ may be distinguished as the *electric vibrations*, those determined by equations of type $G_n = 0$ as the *magnetic vibrations*. Some of the roots of the equations have been calculated for the case of a totally reflecting sphere by Sir J. J. Thomson*, who finds that the roots are all complex. The vibrations for the space inside a totally reflecting sphere have been discussed by Prof. J. W. Nicholson†, those for the space between two concentric spheres by Sir J. J. Thomson‡, Sir Joseph Larmor§, Prof. H. M. Macdonald‖ and A. Lampa¶.

P. Debye, who has calculated some of the roots for a case of a dielectric sphere, finds that the roots are complex and of two types. When the index of refraction is large, the imaginary part of a root of the first type varies very little with the index of refraction N and approaches a limit different from zero when $N \to \infty$. If on the other hand ρ is a root of the second type, $N\rho$ tends to a finite real limit, viz. a root of $\psi_n(N\rho) = 0$, when $N \to \infty$ and so the imaginary part of a root of this type must be very small when N is large.

The vibrations belonging to the space outside a sphere must be in all cases damped on account of the loss of energy by radiation; when the refractive indices of the outside medium and sphere are very nearly equal, they are clearly very strongly damped; thus it is only when the refractive index is large that some of them are durable. It is doubtful whether a substance exists which has a large refractive index and does not absorb light to a marked extent.

* *Proc. London Math. Soc.* Ser. 1, Vol. 15 (1884), p. 197; *Recent Researches*, p. 361.

† *Phil. Mag.* 1906, p. 703.

‡ *Recent Researches*, p. 373.

§ *Proc. London Math. Soc.* Ser. 1, Vol. 26 (1894), p. 119.

‖ *Electric Waves*, Chapters 6–7.

¶ *Wien. Ber.* 112 (1903), p. 37.

It will be seen later that the characteristic vibrations play an important part in determining the size of the sphere on which the pressure of a given type of incident radiation has a maximum value.

Prof. Love* has used the solutions corresponding to the characteristic vibrations to discuss the mode of decay of an arbitrary initial disturbance. He makes use, in fact, of the functions $\zeta_n(kr)$, where k is one of the roots of one of the equations $G_n = 0$, $L_n = 0$; only the sphere is treated as a perfect conductor.

This method can easily be extended so as to provide us with a method of discussing the problem of the scattering of an arbitrary primary disturbance by a spherical obstacle. In this method we assume that the total disturbance outside the sphere can be represented by

$$\left.\begin{aligned} rU &= \sum_{n=0}^{\infty} \sum_{m=0}^{\infty} \sum_{p} A_{n,m,p} \psi_n(k_p r) P_n^m(\cos\theta) \cos m(\phi - \phi_0) \\ rV &= \sum_{n=0}^{\infty} \sum_{m=0}^{\infty} \sum_{p} B_{n,m,p} \psi_n(k_p' r) P_n^m(\cos\theta) \sin m(\phi - \phi_0) \end{aligned}\right\} \quad \ldots\ldots\ldots(\omega),$$

where the k_p's are roots of one of the equations of the type

$$\left.\begin{aligned} \eta\psi_n'(ha)\psi_n(ka) - \nu\psi_n'(ka)\psi_n(ha) &= 0 \\ \nu\psi_n'(ha)\psi_n(ka) - \eta\psi_n'(ka)\psi_n(ha) &= 0 \end{aligned}\right\} \ldots(100).$$

The coefficients must then be chosen so that this total disturbance has the same character as the primary disturbance at its singularities outside the sphere and at an infinite distance, taking into account of course the presence of diverging waves from the spherical obstacle the effect of which is, however, negligible at infinity.

The field inside the sphere is represented by equations similar to (ω), one with h_p written in place of k_p. The boundary conditions are satisfied in virtue of (100).

If a series of type (ω) should fail to represent the disturbance outside the sphere, it may be necessary to add terms corresponding to the free characteristic vibrations. These are of the form (ω) with the function ζ_n written in place of ψ_n and numbers k_n determined by equations of type $L_n = 0$, $G_n = 0$.

* *Proc. London Math. Soc.* Ser. 2, Vol. 2 (1904), p. 88.

§ 19. The case of a very small obstacle.

When a, the radius of the sphere, is very small compared with the wave-length λ of the incident radiation, we may treat ka and ha as small quantities. We may then obtain some idea of the relative magnitudes of the different coefficients in our series by using the expansions (60) and (61).

It is easy to see that the values of A_n, B_n decrease very rapidly in absolute magnitude as n increases. The disturbance radiated from the sphere can consequently be represented approximately by superposing a small number of partial waves, the effect of the others being negligible.

Remembering that $\nu h = \eta k$, we find that when H is finite

$$\left.\begin{aligned} H_n &\sim \frac{\nu k^n h^{n+1}(k^2-h^2)\,a^{2n+3}}{(2n+3)\,.\,1^2\,.\,3^2\ldots(2n+1)^2}, \quad G_n \sim i\eta\left(\frac{h}{k}\right)^n e^{-ika} \\ E_n &\sim \frac{(n+1)\,k^n h^{n-1}\eta\,(k^2-h^2)\,a^{2n+1}}{1^2\,.\,3^2\ldots(2n+1)^2}, \quad F_n \sim -i \\ L_n &\sim i\nu h^n k^{-n-2}\frac{(n+1)\,k^2+nh^2}{2n+1}e^{-ika} \end{aligned}\right\}\ldots(101),$$

and it is easy to see that all our series converge.

It appears from these expressions that the nth magnetic wave, i.e. the disturbance due to the nth term in the expansion for U_1, is of the same order of magnitude as the $(n+1)$th electric wave, i.e. the disturbance due to the $(n+1)$th term in the expansion of V_1. This is in sharp contrast with the result obtained by Sir J. J. Thomson for the case of the totally reflecting sphere wherein the nth electric wave and the nth magnetic wave are of the same order of magnitude.

The first electric wave is clearly of chief importance and Mie has proposed to call this *Rayleigh's radiation*. We easily find that

$$\left.\begin{aligned} B_1 &\sim +ik\frac{k^2-h^2}{2k^2+h^2}a^3, \quad B_2 \sim \frac{-1}{2\,.\,3^2}\frac{k^3(k^2-h^2)}{3k^2+2h^2}a^5 \\ A_1 &\sim \frac{ik}{30}(k^2-h^2)\,a^5 \end{aligned}\right\}\ldots(102).$$

The following diagrams, which are taken from Mie's paper, indicate the character of the electric lines of force for the first four partial vibrations of each type. For the magnetic waves

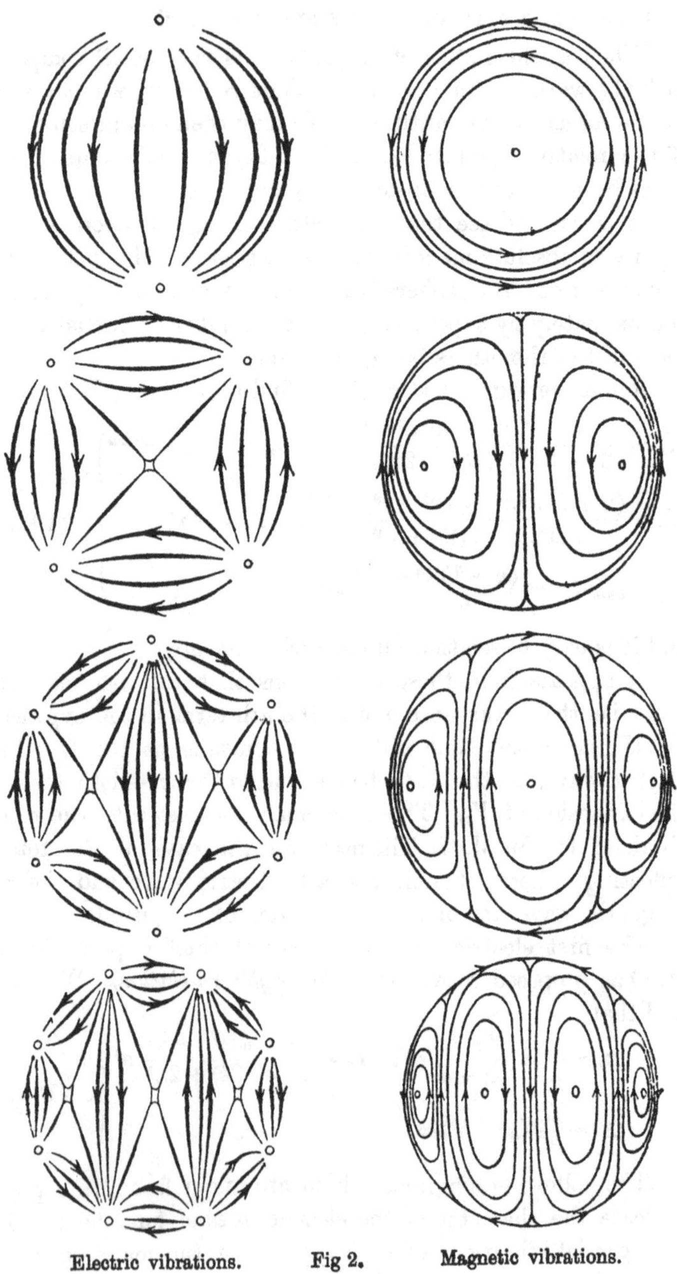

Electric vibrations. Fig 2. Magnetic vibrations.

$E_r = 0$ and so the electric lines of force are spherical curves. In the case of the electric waves the lines of force lie on certain cones and the diagrams represent the intersections of a sphere with these cones, the vertices of the cones being at the centre of the sphere*.

§ 20. **Polarisation of the scattered light.**

Let us now look for cases when the light scattered by the sphere is linearly polarised. It is easy to see that E_θ and M_ϕ both vanish when $\frac{\partial V_1}{\partial \theta} = 0$ and $\frac{\partial U_1}{\partial \phi} = 0$. These conditions are both satisfied by $\phi = 0$, i.e. when the observer looks in a direction at right angles to the electric vibration in the incident wave (Fig. 3).

It appears from the figure that the component of the electric vector of the scattered light, which is at right angles to the direction in which the observer is looking, is parallel to the electric vector in the incident wave. In a similar way it is found that E_ϕ and M_θ both vanish when $\phi = \pm\frac{\pi}{2}$, i.e. when the observer looks in a direction at right angles to the magnetic vibration in the incident wave. The magnetic vibrations in the incident and scattered waves are now found to be parallel.

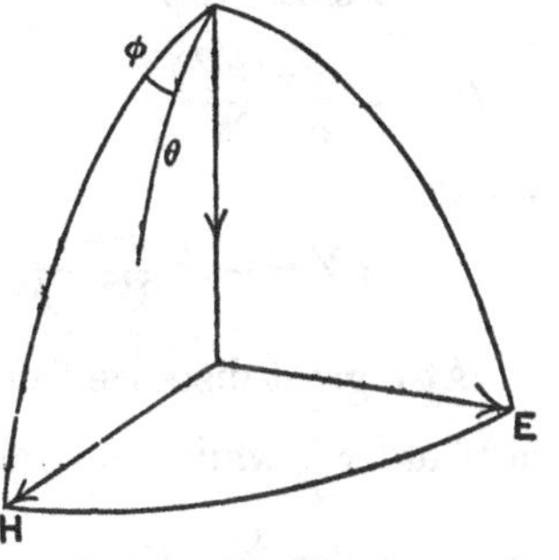

Fig. 3.

The experiments of Steubing† with different kinds of colloidal gold solutions have shown that when the solution is illuminated with polarised light and viewed in the manner described, there is always a small quantity of unpolarised light sent out from the particles, but the greater portion of the scattered light is polarised in the way the theory requires. The slight disagreement between the theory and observations is attributed to the fact that the metallic particles are probably

* The author is indebted to the publisher of the *Annalen der Physik*, Herr Johann Ambrosius Barth, for permission to reproduce the figures on pp. 52, 59 and 64.

† *Dissertation*, Greifswald (1908); *Ann. d. Phys.* Vol. 26 (1908), p. 329.

not all spheres*, that some may have developed into crystals perhaps of octahedral form. The mathematical theory of the scattering of waves has not yet been fully developed. The problem is, however, one of great importance in meteorological optics. The case of a regular distribution of atoms or molecules has recently been brought into prominence† by experimental work on the scattering of Röntgen rays by a crystal‡. Approximate mathematical theories have been given by several writers§.

§ 21. **Intensity of the scattered light.**

When Rayleigh's radiation alone is considered, we have

$$\left.\begin{aligned} H_r &= 0, & E_r &= \frac{1}{\nu}\left[\frac{\partial^2}{\partial r^2}(rV) + k^2 rV\right] \\ H_\theta &= \frac{ik}{r\sin\theta}\frac{\partial(rV)}{\partial\phi}, & E_\theta &= \frac{1}{\nu r}\frac{\partial^2(rV)}{\partial r\partial\theta} \\ H_\phi &= -\frac{ik}{r}\frac{\partial(rV)}{\partial\theta}, & E_\phi &= \frac{1}{\nu r\sin\theta}\frac{\partial^2(rV)}{\partial r\partial\phi} \end{aligned}\right\} \text{...(103)},$$

where

$$rV = -ik\frac{k^2-h^2}{2k^2+h^2}a^3e^{-ikr}\left(1-\frac{i}{kr}\right)\sin\theta\sin\phi \text{...(104)}.$$

At a great distance from the origin the radial electric force is of order $\frac{1}{r^2}$ while the transverse electric and magnetic forces are of order $\frac{1}{r}$. Hence the intensity of the scattered light diminishes ultimately according to the inverse square law when points on the same radius are considered. It also varies as the square of the volume of the particle.

* This remark is made by both Maxwell Garnett and Mie.

† It had previously been considered by Lord Rayleigh, "On the influence of obstacles arranged in rectangular order on the properties of a medium," *Phil. Mag.* (5), Vol. 34 (1892), p. 481; *Scientific Papers*, Vol. 3, p. 19; and by T. H. Havelock, *Proc. Roy. Soc.* A, Vol. 77 (1906), p. 170.

‡ Laue, Friedrich, und Knipping, *Sitzungsber. der Königl. Bayerischen Akad. d. Wiss.* June 1912.

§ See, for instance, W. L. Bragg, *Proc. Camb. Phil. Soc.* Vol. 17 (1913), p. 43; *Proc. Roy. Soc.* A, Vol. 88 (1913), p. 428; M. Laue, *Münchener Ber* (1912), p. 363; *Ann. d. Phys.* Bd. 41 (1913), p. 989, Bd. 42 (1913), p. 397; P. P. Ewald, *Phys. Zeitschr.* (1913), p. 465; L. S. Ornstein, *Amsterdam Proc.* (1913); M. Born u. T. v. Karman, *Phys. Zeitschr.* (1912), p. 297.

An approximate formula for the intensity is

$$I = k^4 a^6 \left(\frac{k^2 - h^2}{2k^2 + h^2}\right)^2 (\cos^2\theta \sin^2\phi + \cos^2\phi) \quad \text{...(105)},$$

the intensity of the incident radiation being $\frac{1}{2}$.

If $\sigma = 0$ for the medium outside the sphere, the quantity k is inversely proportional to the wave-length λ of the incident radiation. Hence when h is large compared with k, we have Lord Rayleigh's result that the intensity of the scattered light varies inversely as the fourth power of the wave-length. The short waves are on this account scattered far more profusely than the long ones and so we have an explanation of the blue colour of the sky.

The above formula for the intensity of Rayleigh's radiation indicates that there is no light of this type in a direction for which $\theta = \phi = \frac{\pi}{2}$, i.e. when the observer is looking in a direction parallel to the direction of the electric vibration in the incident wave. To obtain an expression for the intensity of the light sent out in this direction we must take into account the second electric wave and the first magnetic wave. Referring back to the expressions for B_2 and A_1, we find that when σ is neglected both inside and outside the sphere, the intensity of the scattered radiation varies inversely as the eighth power of the wave-length*. This corresponds to Tyndall's "residual blue" which is purer than the blue seen under other conditions.

§ 22. **The absorption of light by a spherical obstacle.** The total energy absorbed from the incident radiation by a single particle consists of two parts; first of all the energy scattered and secondly the energy which flows into the particle and is transformed into heat or chemical energy. Both these quantities may be calculated by the following method which is a simplification of the one given originally by Mie.

The flow of energy across unit area of a very large sphere concentric with the spherical particle takes place at a rate measured by the radial component of Poynting's vector, i.e. $E_\theta' H_\phi' - E_\phi' H_\theta'$.

* Lord Rayleigh, *Phil. Mag.* Vol. 12 (1881), p. 81.

Now if E, $\bar{E}$ and H, $\bar{H}$ are conjugate complex quantities, this component is equal to

$$\tfrac{1}{4}[(E_\theta e^{i\omega t}+\bar{E}_\theta e^{-i\omega t})(H_\phi e^{i\omega t}+\bar{H}_\phi e^{-i\omega t}) \\ -(E_\phi e^{i\omega t}+\bar{E}_\phi e^{-i\omega t})(H_\theta e^{i\omega t}+\bar{H}_\theta e^{-i\omega t})].$$

Integrating with regard to t so as to obtain the mean value of this quantity over a period, we obtain an expression which may be written in the form†

$$S=\frac{1}{8i\nu}[M_\theta\bar{M}_\phi-\bar{M}_\theta M_\phi-M_\theta{}^*\bar{M}_\phi{}^*+\bar{M}_\theta{}^*M_\phi{}^*],$$

where M, M^* are the values of M corresponding to the signs $+$, $-$ respectively in (86) and $\bar{M}$, $\bar{M}^*$ are derived from them by changing the sign of i.

The function Ω for the outer space is the sum of Ω_0 and Ω_1, hence we may write

$$\left.\begin{aligned} r\Omega &= \sum_{n=1}^{\infty} P_n{}^1(\cos\theta)[u_n e^{i\phi}+v_n e^{i\phi}+w_n e^{-i\phi}] \\ r\bar{\Omega} &= \sum_{n=1}^{\infty} P_n{}^1(\cos\theta)[\bar{u}_n e^{-i\phi}+\bar{v}_n e^{-i\phi}+\bar{w}_n e^{i\phi}] \\ r\Omega^* &= \sum_{n=1}^{\infty} P_n{}^1(\cos\theta)[u_n e^{-i\phi}+v_n e^{-i\phi}+w_n e^{i\phi}] \\ r\Omega^* &= \sum_{n=1}^{\infty} P_n{}^1(\cos\theta)[\bar{u}_n e^{i\phi}+\bar{v}_n e^{-i\phi}+\bar{w}_n e^{i\phi}] \end{aligned}\right\}\ldots(106),$$

where

$$u_n=\frac{1}{k^2}i^{n-1}\frac{2n+1}{n(n+1)}\psi_n(kr),\qquad v_n=\frac{1}{2}(A_n+B_n)\zeta_n(kr),$$

$$w_n=\frac{1}{2}(A_n-B_n)\zeta_n(kr).$$

When the expression S is integrated over the spherical surface, the result can be expressed in the form

$$I_0+I_1-I_{01},$$

where I_0 depends only on the incident radiation, I_1 depends only on the scattered radiation and represents the amount of energy absorbed by scattering, I_{01} depends on both types of radiation and represents the total absorption of energy from

† We assume now that $\sigma=0$ outside the sphere.

the incident wave. The sum of the three terms with its sign changed represents the amount of energy that flows into the sphere and is truly absorbed by it.

Performing the integration with regard to ϕ and collecting the different terms together, we find that the surface integral can be written in the form

$$\frac{\pi}{2i\nu}\sum_{n=1}^{\infty}\sum_{m=1}^{\infty}\left[\int_0^\pi\left(P_n^1P_m^1+\sin^2\theta\frac{dP_n^1}{d\theta}\frac{dP_m^1}{d\theta}\right)\frac{d\theta}{\sin\theta}\right.$$

$$\times\left\{k(u_n+v_n)\left(\frac{d\bar{u}_m}{dr}+\frac{d\bar{v}_m}{dr}\right)+kw_n\frac{d\bar{w}_m}{dr}-\bar{k}(\bar{u}_n+\bar{v}_n)\left(\frac{du_m}{dr}+\frac{dv_m}{dr}\right)\right.$$

$$\left.-\bar{k}\bar{w}_n\frac{dw_m}{dr}\right\}-i\int_0^\pi\left(\frac{dP_n^1}{d\theta}P_m^1+\frac{dP_m^1}{d\theta}P_n^1\right)d\theta$$

$$\times\left\{\left(\frac{du_n}{dr}+\frac{dv_n}{dr}\right)\left(\frac{d\bar{u}_m}{dr}+\frac{d\bar{v}_m}{dr}\right)-\frac{dw_n}{dr}\frac{d\bar{w}_m}{dr}\right.$$

$$\left.\left.+k\bar{k}(u_n+v_n)(\bar{u}_m+\bar{v}_m)-k\bar{k}w_n\bar{w}_m\right\}\right].$$

This expression can be simplified with the aid of the relations

$$\int_0^\pi\left(\frac{dP_n^1}{d\theta}P_m^1+\frac{dP_m^1}{d\theta}P_n^1\right)d\theta=0 \quad \text{.........(107)},$$

$$\left.\begin{aligned}\int_0^\pi\left(P_n^1P_m^1+\sin^2\theta\frac{dP_n^1}{d\theta}\frac{dP_m^1}{d\theta}\right)\frac{d\theta}{\sin\theta}&=0 \qquad m\neq n\\ &=\frac{2n^2(n+1)^2}{2n+1} \qquad m=n\end{aligned}\right\}\text{...(108)}.$$

The first of these is evident since the integrand is the differential coefficient of a function which vanishes at both limits. To prove the second we make use of the formulae

$$\left.\begin{aligned}\sin\theta\frac{d}{d\theta}P_n^1(\cos\theta)&=-\cos\theta\,.\,P_n^1(\cos\theta)\\&\qquad-n(n+1)\sin\theta\,.\,P_n(\cos\theta)\\ P_n^1(\cos\theta)&=-\frac{d}{d\theta}P_n(\cos\theta)\\ \frac{1+\cos^2\theta}{\sin\theta}P_n^1P_m^1&+n(n+1)\cos\theta\,.\,P_nP_m^1\\&\qquad+m(m+1)\cos\theta\,.\,P_mP_n^1\\&=-\frac{d}{d\theta}[\cos\theta\,.\,P_n^1P_m^1]\end{aligned}\right\}\text{...(109)}.$$

The integral is thus equivalent to

$$n(n+1)\,m(m+1)\int_0^\pi P_n(\cos\theta)\,P_m(\cos\theta)\sin\theta\,d\theta,$$

and so has the value we have assigned to it*.

Our surface integral now reduces to

$$\frac{\pi}{iv}\sum_{n=1}^{\infty}\left\{k(u_n+v_n)\left(\frac{d\bar{u}_n}{dr}+\frac{d\bar{v}_n}{dr}\right)+kw_n\frac{d\bar{w}_n}{dr}\right.$$
$$\left.-\bar{k}(\bar{u}_n+\bar{v}_n)\left(\frac{du_n}{dr}+\frac{dv_n}{dr}\right)-\bar{k}\bar{w}_n\frac{dw_n}{dr}\right\}\frac{n^2(n+1)^2}{2n+1}.$$

Since $\sigma=0$ outside the sphere, we have $\bar{k}=k$. Also when r is very large we may use the approximations

$$\left.\begin{aligned}
u_n &\sim \cos\left\{kr-\overline{n+1}\,\frac{\pi}{2}\right\}k^{-2}i^{n-1}\frac{2n+1}{n(n+1)}\\
v_n &\sim \tfrac{1}{2}(A_n+B_n)\,i^{n+1}e^{-ikr}\\
w_n &\sim \tfrac{1}{2}(A_n-B_n)\,i^{n+1}e^{-ikr}\\
\bar{u}_n &\sim \cos\left\{kr-\overline{n+1}\,\frac{\pi}{2}\right\}k^{-2}(-i)^{n-1}\frac{2n+1}{n(n+1)}\\
\bar{v}_n &\sim \tfrac{1}{2}(\bar{A}_n+\bar{B}_n)\,e^{ikr}(-i)^{n+1}\\
\bar{w}_n &\sim \tfrac{1}{2}(\bar{A}_n-\bar{B}_n)\,e^{ikr}(-i)^{n+1}
\end{aligned}\right\}\ldots(110).$$

We thus find that $I_0=0$ and

$$I_0=\frac{\pi}{2iv}\sum_{n=1}^{\infty}n(n+1)\left[i^n(\bar{A}_n+\bar{B}_n)-(-i)^n(A_n+B_n)\right]\ldots(111),$$

$$I_1=\frac{\pi k^2}{\nu}\sum_{n=1}^{\infty}\frac{n^2(n+1)^2}{2n+1}(A_n\bar{A}_n+B_n\bar{B}_n)\;\ldots\ldots\ldots\ldots(112).$$

When $\nu=1$ these expressions are just half those given by Mie on p. 436 of his memoir. The reason why they must be doubled is that our expressions for the incident light give an intensity $\frac{1}{2}$.

In the case of a solution containing N spherical particles per unit volume the total absorption coefficient is thus

$$A=\frac{\pi N}{iv}\sum_{n=1}^{\infty}n(n+1)\left[i^n(\bar{A}_n+\bar{B}_n)-(-i)^n(A_n+B_n)\right]\ldots(113).$$

It is interesting to study the variation of this quantity with the wave-length for particles of different sizes. Mie has drawn

* Cf. L. Lorenz, *Oeuvres scientifiques*, p. 526.

the absorption curves for particles of gold varying in size from 20 $\mu\mu$ to 180 $\mu\mu$.

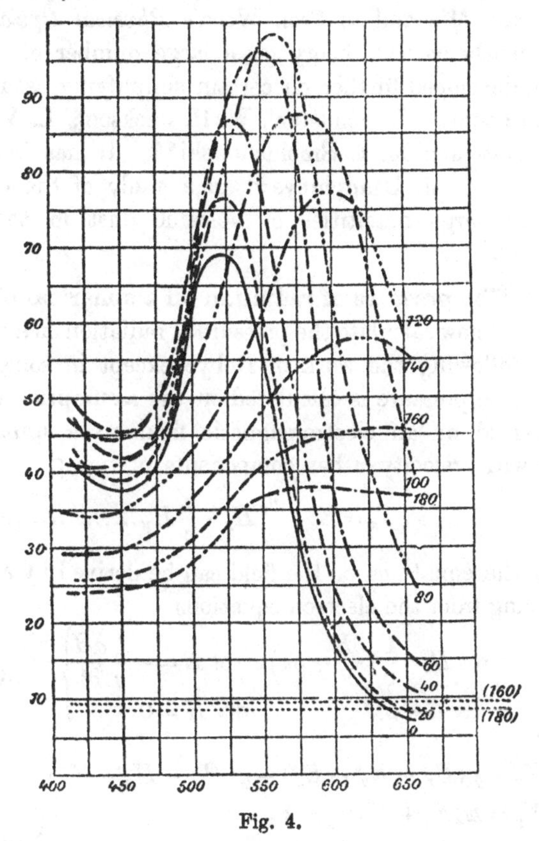

Fig. 4.

It will be seen that for ruby red gold solutions containing very fine particles there is an absorption maximum in the green corresponding to a wave-length of about 525 $\mu\mu$.

Mie has also drawn curves showing the *pure* absorption in colloidal gold solutions.

The colours of silver particles in colloidal silver solutions have been discussed with the aid of the mathematical theory by E. Müller*. The particles of a silver solution show beautiful colour phenomena, all colours of the spectrum from the extreme

* *Ann. d. Phys.* Bd. 35 (1911).

blue to the extreme red being present*. For other applications of the theory we must refer to the papers of Lord Rayleigh, Garnett and Mie and to Prof. Wood's *Physical Optics*.

The mathematical theory for a large number of particles has been developed further on certain simplifying assumptions by F. Hasenöhrl†, A. Schuster‡, W. H. Jackson§, L. V. King‖, A. Einstein¶ and M. v. Smoluchowski**. It has been used recently by W. J. Humphreys†† in a study of the effect on climate of large quantities of volcanic dust in the upper atmosphere.

§ 23. **The pressure of radiation on a spherical obstacle.** We shall now calculate the pressure of radiation on a spherical obstacle, following the work of Debye except in some of the details. The pressure is calculated on the assumption that the force exerted by an electromagnetic field on a unit charge moving with velocity v has components of type‡‡

$$F_x = E_x + \frac{v_y}{c} H_z - \frac{v_z}{c} H_y \quad \text{..............(114)},$$

and that the equations of the field can be derived by a process of averaging from the electron equations

$$\left.\begin{array}{ll} \text{rot}\, H = \dfrac{1}{c}\left(\dfrac{\partial E}{\partial t} + \rho v\right), & \text{rot}\, E = -\dfrac{1}{c}\dfrac{\partial H}{\partial t} \\ \text{div}\, E = \rho, & \text{div}\, H = 0 \end{array}\right\} \quad \text{......(115)}.$$

Now if

$$\left.\begin{array}{l} X_x = \tfrac{1}{2}(E_x^2 - E_y^2 - E_z^2) + \tfrac{1}{2}(H_x^2 - H_y^2 - H_z^2) \\ X_y = E_x E_y + H_x H_y = Y_x \\ X_z = E_x E_z + H_x H_z = Z_x \\ S_x = c(E_y H_z - E_z H_y) \end{array}\right\} \quad \text{...(116)},$$

* A coloured reproduction of an ultramicroscopic picture of a silver solution is given by H. Siedentopf, *Ber. d. Deutsch. Phys. Ges.* (1910), p. 6.

† *Wien. Berichte* (1902). ‡ *Astrophysical Journal*, Vol. 21 (1908).

§ *Bull. of the Amer. Math. Soc.* Vol. 16 (1910), p. 473.

‖ *Phil. Trans.* A, Vol. 212 (1913), p. 375.

¶ *Ann. d. Phys.* Bd. 33 (1910), p. 1275.

** *Boltzmann Festschrift* (1904), p. 626; *Ann. d. Phys.* Bd. 25 (1908), p. 205; *Phil. Mag.* Vol. 23 (1912).

†† *Bull. of the Mount Weather Observatory* (1913); *Journal of the Franklin Institute*, Aug. (1913).

‡‡ We assume now that $\sigma = 0$, $\epsilon = \mu = 1$ for the external medium, so that $\nu = 1$.

etc., we have identically on account of (115)

$$\rho F_x = \frac{\partial X_x}{\partial x} + \frac{\partial X_y}{\partial y} + \frac{\partial X_z}{\partial z} - \frac{1}{c^2}\frac{\partial S_x}{\partial t} \quad \text{.........(117).}$$

Hence by Green's theorem, if (l, m, n) are the direction cosines of the outward drawn normal to a surface σ enclosing all the charges in the field,

$$\iint (lX_x + mX_y + nX_z)\, d\sigma = \iiint \left(\rho F_x + \frac{1}{c^2}\frac{\partial S_x}{\partial t}\right) dx\, dy\, dz \text{...(118).}$$

The x-component of the force exerted by the electromagnetic field on the obstacle is thus the same as if there were tractions X, Y, Z at each point of the surface and a volume force $-\frac{1}{c^2}\frac{\partial S}{\partial t}$: similarly for the other components.

Now when we integrate with regard to t so as to obtain the mean value of the pressure over a period, the term $\frac{\partial S_x}{\partial t}$ contributes nothing on account of the periodicity of S. The pressure may consequently be calculated from Maxwell's tractions X, Y, Z. In the present case the mean value of the pressure in the direction of the axis of z is derived from the quantity*

$$(lZ_x + mZ_y + nZ_z) = -\tfrac{1}{2}\cos\theta\, [E_\theta'^2 + E_\phi'^2 + H_\theta'^2 + H_\phi'^2 - E_r'^2 - H_r'^2] - \sin\theta\, [E_r'E_\theta' + H_r'H_\theta'].$$

The surface integral of the mean value of this quantity over a period is accordingly

$$-\frac{r^2}{4}\int_0^\pi\int_0^{2\pi} [\cos\theta\, (E_\theta\bar{E}_\theta + E_\phi\bar{E}_\phi + H_\theta\bar{H}_\theta + H_\phi\bar{H}_\phi - E_r\bar{E}_r - H_r\bar{H}_r) + \sin\theta\, (E_r\bar{E}_\theta + \bar{E}_r E_\theta + H_r\bar{H}_\theta + \bar{H}_r H_\theta)] \sin\theta \,.\, d\theta\, d\phi.$$

When the incident radiation only is taken into account, this integral becomes simply

$$-\frac{r^2}{2}\int_0^\pi\int_0^{2\pi} \cos\theta \sin\theta\, d\theta\, d\phi$$

and vanishes completely. Again, it is easy to see that when r is large the components E_r, H_r for the scattered field are small

* We assume now that the surface σ is a sphere whose centre is at the origin. Our expression is easily obtained by writing down the expressions for E_x', E_y', E_z' in terms of E_r', E_θ', E_ϕ'.

compared with the transverse components, a product such as $H_r\bar{H}_\theta$ being of order r^{-3} may be neglected when r is large, and so we have only to consider the integral

$$\mathbf{P}_z \equiv -\frac{r^2}{4}\int_0^\pi\int_0^{2\pi}(E_\theta\bar{E}_\theta+E_\phi\bar{E}_\phi+H_\theta\bar{H}_\theta+H_\phi\bar{H}_\phi)\sin\theta\cos\theta\,d\theta d\phi,$$

where $E=E_0+E_1$, $H=H_0+H_1$. The terms which depend only on the incident field may, moreover, be disregarded.

To evaluate $\mathbf{P}_z$ we need the values of the integrals

$$I_1=\int_0^\pi\left(P_n^1P_m^1+\sin^2\theta\frac{dP_n^1}{d\theta}\frac{dP_m^1}{d\theta}\right)\cot\theta\,.\,d\theta,$$

$$I_2=\int_0^\pi\left(\frac{dP_n^1}{d\theta}P_m^1+\frac{dP_m^1}{d\theta}P_n^1\right)\cos\theta\,.\,d\theta.$$

By using the relations (109) we may transform the first of these into

$$n(n+1)\,m(m+1)\int_0^\pi P_n(\cos\theta)\,P_m(\cos\theta)\sin\theta\cos\theta\,d\theta$$
$$-\int_0^\pi\frac{d}{d\theta}[\cos\theta\,.\,P_n^1\,.\,P_m^1]\cos\theta\,.\,d\theta.$$

Now

$$(2n+1)\cos\theta\,.\,P_n(\cos\theta)=(n+1)\,P_{n+1}(\cos\theta)+nP_{n-1}(\cos\theta) \quad\text{.........(119)},$$

and

$$(2n+1)\cos\theta\,.\,P_n^1(\cos\theta)=nP^1_{n+1}(\cos\theta)+(n+1)\,P^1_{n-1}(\cos\theta) \quad\text{.........(120)};$$

hence when the second integral is integrated by parts we obtain two integrals of the type (76) and so we have finally

$$\left.\begin{aligned} I_1&=0 && m\neq n\pm1\\ &=2\frac{(n-1)^2n(n+1)^2}{(2n-1)(2n+1)} && m=n-1\\ &=2\frac{n^2(n+1)(n+2)^2}{(2n+1)(2n+3)} && m=n+1\end{aligned}\right\}\quad\text{......(121)}.$$

When the second integral I_2 is integrated by parts it becomes

$$\left.\begin{aligned} I_2=\int_0^\pi P_n^1P_m^1\sin\theta d\theta&=0 && m\neq n\\ &=2\,\frac{n(n+1)}{2n+1} && m=n\end{aligned}\right\}\quad\text{......(122)}.$$

Writing $\mathbf{P}_z$ in the form

$$\mathbf{P}_z = -\frac{r^2}{8}\int_0^{\pi}\int_0^{2\pi}(M_\theta \overline{M}_\theta + M_\theta{}^* \overline{M}_\theta{}^* + M_\phi \overline{M}_\phi + M_\phi{}^* \overline{M}_\phi{}^*)\sin\theta\cos\theta\, d\theta\, d\phi$$

and making use of equations (121), (122) and (106), we obtain

$$\mathbf{P}_z = -\pi \sum_{n=1}^{\infty} \frac{n^2(n+1)(n+2)^2}{(2n+1)(2n+3)}\Bigg[\left(\frac{du_{n+1}}{dr}+\frac{dv_{n+1}}{dr}\right)\left(\frac{d\bar{u}_n}{dr}+\frac{d\bar{v}_n}{dr}\right) + \frac{dw_{n+1}}{dr}\frac{d\overline{w}_n}{dr} + k^2(u_{n+1}+v_{n+1})(\bar{u}_n+\bar{v}_n) + k^2 w_{n+1}\overline{w}_n$$
$$+\left(\frac{du_n}{dr}+\frac{dv_n}{dr}\right)\left(\frac{d\bar{u}_{n+1}}{dr}+\frac{d\bar{v}_{n+1}}{dr}\right)+\frac{dw_n}{dr}\frac{d\overline{w}_{n+1}}{dr} + k^2(u_n+v_n)(\bar{u}_{n+1}+\bar{v}_{n+1}) + k^2 w_n \overline{w}_{n+1}\Bigg]$$
$$-ki\pi \sum_{n=1}^{\infty}\frac{n(n+1)}{2n+1}\left[(u_n+v_n)\left(\frac{d\bar{u}_n}{dr}+\frac{d\bar{v}_n}{dr}\right) - (\bar{u}_n+\bar{v}_n)\left(\frac{du_n}{dr}+\frac{dv_n}{dr}\right)+\overline{w}_n\frac{dw_n}{dr} - w_n\frac{d\overline{w}_n}{dr}\right].$$

We now use the asymptotic expressions for u_n, v_n, w_n when r is large and omit the terms that depend only on the u's.

We also write

$$A_n = -i^{n-1}\frac{2n+1}{n(n+1)}\alpha_n, \qquad B_n = -i^{n-1}\frac{2n+1}{n(n+1)}\beta_n,$$

and obtain after some simplifications

$$\mathbf{P}_z = -\frac{\pi}{2}\sum_{n=1}^{\infty}(2n+1)[\alpha_n+\bar{\alpha}_n+\beta_n+\bar{\beta}_n] + \pi k^2 \sum_{n=1}^{\infty}\frac{2n+1}{n(n+1)}(\alpha_n\bar{\beta}_n+\bar{\alpha}_n\beta_n)$$
$$+\pi k^2\sum_{n=1}^{\infty}\frac{n(n+2)}{n+1}[\alpha_n\bar{\alpha}_{n+1}+\bar{\alpha}_n\alpha_{n+1}+\beta_n\bar{\beta}_{n+1}+\bar{\beta}_n\beta_{n+1}] \qquad \text{.........(123).}$$

This expression turns out to be negative, i.e. the pressure acts in the direction in which the incident light is moving. If the constants in the incident light are chosen so that its intensity is unity, the above expression must be doubled. The numerical value of the pressure has been calculated from the above formula by Debye in a number of cases. When the radius of the spherical obstacle is small compared with the wave-length of the incident light, the functions $\psi_n(ka)$, $\zeta_n(ka)$, $\psi_n(ha)$, $\zeta_n(ha)$ occurring in the expressions for α_n, β_n can be expanded in ascending powers of a. This method, however, fails when ka approaches unity and numerical values of the functions must be used.

In the case of a totally reflecting sphere

$$\alpha = \frac{\psi_n'(ka)}{\zeta_n'(ka)}, \qquad \beta = \frac{\psi_n(ka)}{\zeta_n(ka)} \quad \text{.........(124)},$$

and Debye finds that if

$$\rho = ak = \frac{2\pi a}{\lambda},$$

where λ is the wave-length of the incident light, L denotes the light-pressure and $W = \frac{1}{2}\pi a^2$ is the energy of the incident train of waves per unit length of a cylinder circumscribing the spherical obstacle and having its axis parallel to the direction of motion of the waves; then

$$\frac{L}{W} = \frac{14}{3}\rho^4 [1 + \tfrac{1}{21}\rho^2 - \tfrac{1409}{8820}\rho^4 \ldots] \quad \text{.........(125)}.$$

The first term of the series was given by Schwarzschild. The convergence of the series is slow as ρ approaches unity and several terms of the series (123) must be taken into account. By using numerical values of the functions $\psi_n(\rho)$, $\zeta_n(\rho)$ and their derivatives for $n = 1, 2 \ldots 5$, Debye has succeeded in drawing a curve for $\frac{L}{W}$.

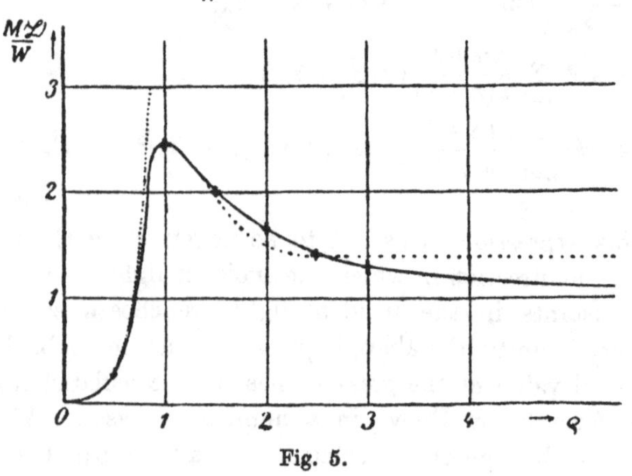

Fig. 5.

It will be seen that the pressure has a maximum value for a certain value of ρ, approximately equal to 1. When ρ is large the ratio $\frac{L}{W}$ approaches asymptotically the value 1.

Debye compares the light pressure so obtained with the gravitational attraction for a spherical particle of specific gravity s under the influence of the sun's radiation. He finds that if G is the gravitational attraction,

$$\frac{L}{G}=\frac{4800}{\lambda s\rho}\cdot\frac{L}{W},$$

and to get a numerical estimate he takes $\lambda = 600\ \mu\mu$, $s = 1$.

It appears that the ratio vanishes both for small and large values of ρ: it has a maximum value of about 20 for $\rho = 1$.

In the case of a dielectric sphere with refractive index n, the expansion corresponding to (125) is

$$\frac{L}{W}=\frac{8}{3}\left(\frac{n^2-1}{n^2+1}\right)^2\rho^4\left[1-\frac{\rho^2}{15}\cdot\frac{n^6-29n^4+34n^2+120}{(n^2+2)(2n^2+3)}-\dots\right] \quad \text{......(126)},$$

and is suitable for calculations only when $n\rho$ is small.

Debye has drawn curves for $\frac{L}{W}$ in the cases $n=\infty$, $n=2$, $n=1{\cdot}5$ and $n=1{\cdot}33$. When $n=2$ the curve appears to have three maxima and two minima between the values $\rho=1$ and $\rho=3$.

The greatest value of $\frac{L}{W}$ is now about 2·6; the following table indicates when the light pressure exceeds the gravitational attraction, the numbers ρ_0 and ρ_1 give the extreme values of ρ belonging to the range in which this is the case.

n	$\left(\frac{L}{G}\right)_{\text{Max.}}$	ρ_0	ρ_1
∞	20	·3	8
2	13	·6	5
1·5	3	·8	—

The maximum light pressure is just balanced by the gravitational action when n is about equal to 1·33, the value for water: for smaller values of n gravitation prevails.

In the case of an absorbing spherical particle, the equation which takes the place of (126) is

$$\frac{L}{W} = 12 \frac{\frac{\sigma}{\omega}}{(\kappa+2)^2 + \frac{\sigma^2}{\omega^2}} \rho \quad \text{..............(127).}$$

When a is small the light pressure and gravitational action are both of order a^3 and their ratio tends to a finite limit, hence for certain types of absorbing material there is no lower limit in the size of a particle below which gravitation exceeds the light pressure. Debye has drawn a curve for $\frac{L}{W}$ for the case of a gold particle and finds that there is a maximum value for $\rho = 1{\cdot}5$ nearly.

The existence of a maximum value for $\frac{L}{W}$ in the cases that have been discussed appears to be due to the fact that the value of ρ for which the maximum occurs is very nearly equal to the real part of the complex value of ρ corresponding to one of the free damped vibrations*. The first electric vibration seems to be of chief importance in determining the position of the maximum.

The determination of the limiting value of $\frac{L}{W}$ for very small wave-lengths, i.e. for large values of ρ, is a matter of some difficulty, it depends on some expressions giving the behaviour of the Bessel functions for large values of n and ρ. These have been found by J. W. Nicholson† and P. Debye‡.

If ρ is real and $n + \frac{1}{2} < \rho$, we have when $\rho \to \infty$

$$\left.\begin{aligned} \zeta_n(\rho) &= \frac{e^{-i\left(\rho f_0 - \frac{\pi}{4}\right)}}{(\sin \tau_0)^{\frac{1}{2}}} \\ \psi_n(\rho) &= \frac{\cos\left(\rho f_0 - \frac{\pi}{4}\right)}{(\sin \tau_0)^{\frac{1}{2}}} \end{aligned}\right\} \quad \text{..............(128),}$$

* Cf. Debye, *loc. cit.*, and the similar remarks for the case of optical resonance by F. Pockels, *Physik. Zeitschr.* Bd. 5 (1904), p. 152.

† *British Association Reports*, Dublin, 1908, p. 595; *Phil. Mag.* Vol. 13 (1906), p. 195; Vol. 14 (1907), p. 697; Vol. 16 (1908), p. 271; Vol. 18 (1909), p. 6.

‡ *Math. Ann.* Bd. 67 (1909), p. 535.

where τ_0 is an angle lying between 0 and $\frac{\pi}{2}$ for which

$$\cos \tau_0 = \frac{n + \frac{1}{2}}{\rho}$$

and $$f_0 = \sin \tau_0 - \tau_0 \cos \tau_0.$$

When $n + \frac{1}{2} > \rho$ and $\rho \to \infty$, we have

$$\left. \begin{aligned} \zeta_n(\rho) &= i \frac{e^{-i\rho f_0}}{(i \sin \tau_0)^{\frac{1}{2}}} \\ \psi_n(\rho) &= \frac{e^{i\rho f_0}}{(i \sin \tau_0)^{\frac{1}{2}}} \end{aligned} \right\} \quad \text{..............(129)},$$

where τ_0 is now the root of the equation

$$\cos \tau_0 = \frac{n + \frac{1}{2}}{\rho}$$

whose imaginary part has a negative sign.

When n and z are very nearly equal the values of $\zeta_n(\rho)$ and $\psi_n(\rho)$ can be made to depend on Airy's integral and are much more complicated; for these we must refer the reader to the original memoirs.

24. Other problems which may be treated with the aid of polar coordinates.

The diffraction of electric waves travelling round the earth is a problem of some importance which has been discussed by H. M. Macdonald*, Lord Rayleigh†, H. Poincaré‡, J. W. Nicholson§ and other writers.

The calculations are very long and depend on the use of the formulae to which we have just referred. Rybcyński|| has recently treated the problem by a method due to March and has taken into account the finite conductivity of the earth. As we have already mentioned this was done by Zenneck and Sommerfeld for the case in which the earth's surface is treated

* *Proc. Roy. Soc.* Vol. 71 (1903), p. 251; Vol. 72 (1904), p. 59; Vol. 90 (1914), p. 50; *Phil. Trans.* A, Vol. 210 (1909), p. 113.

† *Proc. Roy. Soc.* Vol. 72 (1904), p. 40.

‡ *Rend. Palermo* (1910); *Proc. Roy. Soc.* Vol. 72 (1904), p. 42.

§ *Phil. Mag.* Vol. 19 (1910), pp. 276, 435, 516, 757; Vol. 20 (1911), p. 157; Vol. 21 (1911), pp. 62, 281; *Jahrb. d. draht. Telegr.* Bd. 4 (1910), p. 20.

|| *Ann. d. Phys.* Bd. 41 (1913).

as a plane. The results of Nicholson and Poincaré indicate that diffraction round a perfectly conducting surface is not sufficient to explain the apparent bending of the electric waves round the earth's surface. A generally accepted opinion is that the ionisation of the atmosphere by the sun's rays is a very important factor in producing the observed effects*.

The diffraction of a solitary wave or pulse by a spherical obstacle might be discussed with advantage. The evaluation of certain definite integrals involving Bessel functions, however, presents some formidable difficulties which probably account for the fact that the problem does not appear to have been solved.

The scattering of electric waves by a perfectly conducting conical obstacle has been treated very briefly by H. S. Carslaw†.

EXAMPLES.

1. Prove that

$$1+\sum_{1}^{\infty}\frac{2n+1}{n(n+1)}P_n(\cos\theta)\,P_n(\cos a)=\log\left(\sec^2\frac{a}{2}\operatorname{cosec}^2\frac{\theta}{2}\right),\quad 0\leqslant a\leqslant\theta\leqslant\pi.$$

(C. Neumann.)

2. Prove that

$$e^{kiz\cos a}J_0(k\rho\sin a)=\sqrt{\frac{\pi}{2}}\sum_{1}^{\infty}(2n+1)\,i^n\,\frac{J_{n+\frac{1}{2}}(kr)}{\sqrt{kr}}\,P_n(\cos\theta)\,P_n(\cos a).$$

(E. W. Hobson.)

* Cf. the discussion at the British Association meeting, Dundee (1912), and an article by W. H. Eccles in the *Year Book of Wireless Telegraphy* (1913). Some quantitative experiments on long distance telegraphy have been made recently by L. W. Austin, who obtains an empirical relation between the magnitude of the current received and the distance between the two stations, *Bulletin of the Bureau of Standards*, Vol. 7 (1911).

† *Phil. Mag.* Vol. 20 (1910), p. 690.

CHAPTER IV

CYLINDRICAL COORDINATES

§ 25. The wave-equation in Cylindrical Coordinates.

If we put $x = \rho\cos\phi$, $y = \rho\sin\phi$, the wave-equation becomes

$$\frac{\partial^2 u}{\partial \rho^2} + \frac{1}{\rho}\frac{\partial u}{\partial \rho} + \frac{1}{\rho^2}\frac{\partial^2 u}{\partial \phi^2} + \frac{\partial^2 u}{\partial z^2} - \frac{1}{c^2}\frac{\partial^2 u}{\partial t^2} = 0 \quad \text{......(130).}$$

Two particular solutions of this equation are suggested at once by the general solution of § 5: they are*

$$u = \int_0^{2\pi} F\left[z + i\rho\cos\alpha, \quad t - \frac{\rho}{c}\sin\alpha\right] d\alpha \quad \text{......(131),}$$

and
$$u = \int_{-\phi}^{+\phi} F\left[z + i\rho\cos\alpha, \quad t - \frac{\rho}{c}\sin\alpha\right] d\alpha \quad \text{......(132),}$$

respectively. The first of these represents a wave-function which is symmetrical round the axis of z and which reduces to $2\pi F(z, t)$ when $\rho = 0$. It gives us at once the formulae

$$r^n P_n(\mu) = \frac{1}{2\pi}\int_0^{2\pi} (z + i\rho\cos\alpha)^n\, d\alpha \quad \left(\mu = \frac{z}{r}\right) \text{...(133),}$$

$$\frac{1}{r^{n+1}} P_n(\mu) = \frac{1}{2\pi}\int_0^{2\pi} \frac{d\alpha}{(z + i\rho\cos\alpha)^{n+1}} \quad \left(\mu = \frac{z}{r}\right) \text{...(134),}$$

where n is an integer†, and many other interesting formulae may be written down by simply choosing different wave-functions that are symmetrical round the axis.

* The first of these is an obvious generalisation of a formula given by D. Edwardes, *Educational Times*, Oct. (1904).

† The formula is also true under certain limitations when n is not an integer. See Hobson, *Phil. Trans.* A, Vol. 187 (1896).

For instance,

$$\frac{1}{\pi}\int_0^\pi \frac{\log(z+i\rho\cos\theta)}{z+i\rho\cos\theta}\,d\theta=\frac{1}{r}\log\frac{2r^2}{r+z} \quad \text{............(135)},$$

$$\frac{1}{\pi}\int_0^\pi \frac{\log(z+i\rho\cos\theta)-\log z_0}{z-z_0+i\rho\cos\theta}\,d\theta=\frac{1}{R}\log\frac{r+z_0+R}{r+z_0-R}$$

$$(R^2=\rho^2+(z-z_0)^2,\ z>z_0)\text{...(136)},$$

$$\frac{1}{2\pi}\int_0^{2\pi} \frac{\sin k(z+i\rho\cos\alpha)}{z+i\rho\cos\alpha}\,e^{ik\rho\sin\alpha}\,d\alpha=\frac{\sin kr}{r} \quad \text{...(137)}.$$

There is another general formula for a wave-function symmetrical about an axis, viz.

$$u=\int_{-\infty}^{\infty} F\left[t-\frac{\rho}{c}\cosh\alpha,\quad z-\rho\sinh\alpha\right]d\alpha \quad \text{...(138)},$$

where the function F is of such a nature that

$$\frac{1}{c}\sinh\alpha\frac{\partial F}{\partial t}+\cosh\alpha\frac{\partial F}{\partial z}$$

vanishes when $\alpha=\pm\infty$. As a particular instance of this we have the function*

$$\Omega=\frac{1}{2\pi}\int_0^\infty F\left(t-\frac{\rho}{c}\cosh\alpha\right)d\alpha \quad \text{.........(139)},$$

which may be regarded as the cylindrical wave-potential for a line source of strength $F(t)$ along the axis of z.

A peculiarity† of the two-dimensional propagation of waves is the existence of a "tail" to the disturbance when $F(t)$ is zero for $t<0$ and $t>\tau$, for if $t>\tau+\frac{\rho}{c}$ we have

$$\Omega=\frac{1}{2\pi}\int_s^\infty F\left(t-\frac{\rho}{c}\cosh\alpha\right)d\alpha \quad \text{.........(140)},$$

where $\rho\cosh s=c(t-\tau)$. It is clear that this expression for Ω does not generally vanish. The wave-function (139) is thus essentially different from Euler's wave-potential for a point source, viz.

$$\Omega=\frac{1}{r}F\left(t-\frac{r}{c}\right),$$

* Cf. H. Lamb, *Hydrodynamics*, pp. 281, 500; V. Volterra, *Acta Math.* t. 18; Levi-Cività, *Nuovo Cimento* (4), t. 6 (1897). The formula is a particular case of a more general one given by Dr Hobson in 1891.

† This was discussed by O. Heaviside, *Phil. Mag.* (5), t. 26 (1888); *Electrical Papers*, Vol. 2. See also Lamb, *Hydrodynamics*, p. 282.

for in this case Ω is zero for $t > \tau + \frac{r}{c}$, provided the source is only active when $0 < t < \tau$.

V. Volterra* has obtained a number of elementary wave-functions of the form

$$u = t^n F\left(\frac{ct}{\rho}\right) = t^n F(s) \quad \text{.............(141).}$$

He finds that F must satisfy the differential equation

$$s^2(1-s^2)\frac{d^2F}{ds^2} + s(2n - s^2)\frac{dF}{ds} + n(n-1)F = 0 \quad \text{...(142).}$$

If we try to solve (130) by means of a function of the form

$$u = W\rho^m \cos m(\phi - \phi_0),$$

where W is independent of ϕ, we find that W must satisfy the equation

$$\frac{\partial^2 W}{\partial \rho^2} + \frac{2m+1}{\rho}\frac{\partial W}{\partial \rho} + \frac{\partial^2 W}{\partial z^2} - \frac{1}{c^2}\frac{\partial^2 W}{\partial t^2} = 0 \quad \text{......(143).}$$

Solutions of this which are independent of z may be derived from the following formulae†, in which $m > -\frac{1}{2}$,

$$W = \int_0^\pi f\left(t - \frac{\rho}{c}\cos\alpha\right)\sin^{2m}\alpha \,.\, d\alpha \quad \text{.........(144),}$$

$$W = \int_0^\infty F\left(t \pm \frac{\rho}{c}\cosh\eta\right)\sinh^{2m}\eta \,.\, d\eta \quad \text{......(145).}$$

There are, of course, certain limitations concerning the behaviour of $F(t)$ at infinity. The first formula enables us to determine the value of W when its value is known for $\rho = 0$.

§ 26. Elementary solutions of $\Delta u + k^2 u = 0$.

The differential equation

$$\frac{\partial^2 u}{\partial \rho^2} + \frac{1}{\rho}\frac{\partial u}{\partial \rho} + \frac{1}{\rho^2}\frac{\partial^2 u}{\partial \phi^2} + \frac{\partial^2 u}{\partial z^2} + k^2 u = 0 \quad \text{.........(146)}$$

possesses elementary solutions of the types

$$J_m(\rho\sqrt{k^2 + h^2})\, e^{\pm hz} \cos m(\phi - \phi_0) \quad \text{......(147),}$$

$$K_m(\rho\sqrt{\lambda^2 - k^2})\, e^{\pm i\lambda z} \cos m(\phi - \phi_0) \quad \text{......(148),}$$

* *Acta Math.* Vol. 18.

† E. W. Hobson, *Proc. London Math. Soc.* Vol. 22 (1891), p. 431. The first solution is due to Poisson.

where λ, h and m are real or complex arbitrary constants. The first solution may of course be generalised into

$$u=\int_a^\infty J_m(\rho\sqrt{k^2+h^2})\, e^{\pm hz} f(h)\, h\,dh \quad \text{.........(149)},$$

and a similar remark applies to the second. If we wish to express a given wave-function in the first of these forms, the following inversion formula due to Hankel is particularly useful*.

$$\left.\begin{aligned} \text{If} \qquad F(x) &= \int_0^\infty J_m(xt) f(t)\, t\,dt \\ \text{then} \qquad f(t) &= \int_0^\infty J_m(xt) F(x)\, x\,dx \end{aligned}\right\} \quad \text{............(150)}.$$

Let us use this formula to express $\frac{1}{r}e^{ikr}$ in the form (149) when $m=0$. Since the representation should be valid for $z=0$, we find on putting $k^2+h^2=\lambda^2$, $a=k$, that

$$\frac{e^{ik\rho}}{\rho}=\int_0^\infty J_0(\lambda\rho) f(h)\, \lambda\, d\lambda;$$

$$\therefore \quad f(h)=\int_0^\infty J_0(\lambda\rho)\, e^{ik\rho}\, d\rho = \frac{1}{\sqrt{\lambda^2-k^2}}, \quad k>0.$$

Hence we obtain Sommerfeld's formula†

$$\frac{1}{r}e^{ikr}=\int_0^\infty e^{\pm z\sqrt{\lambda^2-k^2}} J_0(\lambda\rho)\frac{\lambda\, d\lambda}{\sqrt{\lambda^2-k^2}} \quad \text{......(151)},$$

the upper or lower sign being taken according as $z \lessgtr 0$. A more complete proof is obtained by applying Hankel's inversion formula to the equation

$$\left.\begin{aligned} \int_0^\infty J_0(\lambda\rho)\, e^{ikr}\frac{\rho\, d\rho}{r} &= \frac{e^{-|z|\sqrt{\lambda^2-k^2}}}{\sqrt{\lambda^2-k^2}} \qquad \lambda^2>k^2 \\ &= \frac{ie^{i|z|\sqrt{k^2-\lambda^2}}}{\sqrt{k^2-\lambda^2}} \qquad \lambda^2<k^2 \end{aligned}\right\} \quad \text{...(152)},$$

which is established in Prof. Lamb's paper.

* See Gray and Mathews, *Treatise on Bessel Functions* (1895), p. 80; N. Nielsen, *Handbuch der Theorie der Cylinderfunktionen* (1904), p. 366.

† *Ann. d. Physik*, Bd. 28 (1909), p. 683. Some analogous formulae are given by H. Lamb, *Proc. London Math. Soc.* Ser. 2, Vol. 7 (1909), p. 140; O. Heaviside, *Electrical Papers*, Vol. 2, p. 478; N. Sonin, *Math. Ann.* Bd. 16.

A few more formulae will now be written down to illustrate the method of generalisation by integration with regard to a variable parameter

$$K_0(\rho\sqrt{\lambda^2-k^2})\cos\lambda(z-b)=\tfrac{1}{2}\int_{-\infty}^{\infty}\frac{e^{ikR}}{R}\cos\lambda(a-b)\,da$$

$$(\lambda^2>k^2,\ R^2=\rho^2+(z-a)^2).$$

If m is zero or a positive integer and n is a positive integer,

$$\frac{1}{r^{n+1}}P_n^m(\cos\theta)=\frac{1}{\Gamma(n-m+1)}\int_0^{\infty}e^{-\lambda z}J_m(\lambda\rho)\lambda^n d\lambda,\qquad z>0,$$

$$\frac{1}{r^{n+1}}Q_n^m(\cos\theta)=\frac{\pi}{2\Gamma(n-m+1)}\int_0^{\infty}e^{-\lambda z}Y_m(\lambda\rho)\lambda^n d\lambda,\qquad z>0.$$

(Hobson.)

§ 27. **The propagation of electric waves on a semi-infinite solid bounded by a plane surface*.**

In this problem the surface of the earth is regarded as an infinite plane and the waves are supposed to be generated by an antenna, of which one portion is vertical and the other horizontal.

Let us assume that the electric and magnetic forces are the real parts of vectors of the form $Ee^{-i\omega t}$, $He^{-i\omega t}$ respectively, then if $M=H\pm i\nu E$, where $\nu^2=\dfrac{\epsilon\omega+i\sigma}{\mu\omega}$, we may satisfy Maxwell's equations by putting

$$M=\text{rot}\,\Pi\pm\frac{1}{k}\,\text{grad div}\,\Pi\pm k\Pi \quad\text{.........(153)},$$

where $\quad \Delta\Pi+k^2\Pi=0,\qquad c^2k^2=\epsilon\mu\omega^2+i\mu\omega\sigma$ ......(154).

To imitate the action of the antenna we shall place two vibrating doublets at a point at distance a from the plane. If one of these vibrates vertically and the other horizontally, we may put for the primary radiation $\Pi_0=(P_x,0,P_z)$, where

$$P_x=-B\frac{e^{ikR}}{R},\quad P_z=A\frac{e^{ikR}}{R},\quad R^2=\rho^2+(z-a)^2\text{...(155)},$$

and the axis of z is vertical. We write B with a negative sign to signify that the horizontal branch of the antenna is drawn in the negative x-direction.

* A. Sommerfeld, *Ann. d. Phys.* Bd. 28 (1909), p. 665; H. v. Hoerschlemann, *Jahrb. d. draht. Teleg.* Bd. 5 (1912), pp. 14, 188; *Dissertation*, Munich (1911).

A convenient expression for Π_0 is obtained by using Sommerfeld's equation

$$\frac{1}{R}e^{ikR} = \int_0^\infty J_0(\lambda)\, e^{l(z-a)}\frac{\lambda d\lambda}{l}, \quad 0<z<a \quad \text{......(156)},$$

where $l = \sqrt{\lambda^2 - k^2}$. Appropriate functions Π_1, Π_2 for the reflected and transmitted disturbances are obtained by putting

$$\left.\begin{aligned} \Pi_1 &= \sum_{n=0}^{\infty} \cos n\phi \int_0^\infty J_n(\lambda\rho)\, e^{-l(z+a)}\, F_n(\lambda)\, d\lambda \\ \Pi_2 &= \sum_{n=0}^{\infty} \cos n\phi \int_0^\infty J_n(\lambda\rho)\, e^{m(z-a)}\, G_n(\lambda)\, d\lambda \end{aligned}\right\} z < a \quad \text{...(157)},$$

where $m = \sqrt{\lambda^2 - h^2}$ and h is the value of k in the second medium. The functions $F_n(\lambda)$, $G_n(\lambda)$ are vectors with components $[f_n(\lambda), 0, \psi_n(\lambda)]$, $[g_n(\lambda), 0, \chi_n(\lambda)]$ respectively.

We can satisfy the condition that the tangential components of the electric and magnetic forces should be continuous at the surface of separation of the two media by putting for $z = 0$

$$\left.\begin{aligned} \Pi_0 + \Pi_1 &= \Pi_2 \\ \text{rot}\,(\Pi_0 + \Pi_1) &= \text{rot}\,\Pi_2 \\ \frac{1}{k^2}\,\text{div}\,(\Pi_0 + \Pi_1) &= \frac{1}{h^2}\,\text{div}\,\Pi_2 \end{aligned}\right\} \quad \text{.........(158)},$$

where μ has been taken to be unity for both media.

Substituting the integral expressions for Π_0, Π_1, Π_2 in these equations and equating to zero the coefficients of functions of type $J_n(\lambda\rho)$ in the resulting integral equation, we obtain the system of equations

$$f_0 e^{-la} - \frac{B}{l}e^{-la} = g_0 e^{-ma}, \qquad f_n e^{-la} = g_n e^{-ma},$$

$$\psi_0 e^{-la} + \frac{A}{l}e^{-la} = \chi_0 e^{-ma}, \qquad \psi_n e^{-la} = \chi_n e^{-ma},$$

$$-lf_0 e^{-la} - Be^{-la} = mg_0 e^{-ma}, \quad -lf_n e^{-la} = mg_n e^{-ma},$$

$$(Ae^{-la} - l\psi_0 e^{-la})\frac{1}{k^2} = \frac{m}{h^2}\chi_0 e^{-ma},$$

$$\left(\frac{B\lambda}{l}e^{-la} - l\psi_1 e^{-la} - \lambda f_0 e^{-la}\right)\frac{1}{k^2} = \frac{1}{h^2}(m\chi_1 e^{-ma} - \lambda g_0 e^{-ma}),$$

$$-\frac{l}{k^2}\psi_n e^{-la} = \frac{m}{h^2}\chi_n e^{-ma},$$

where the last equation has been simplified with the aid of the relations $f_n = g_n = 0\ (n > 0)$ which are a consequence of the previous equations.

Solving these equations we eventually find that if $\Pi_1 = (Q_x,\ 0,\ Q_z)$, $\Pi_2 = (R_x,\ 0,\ R_z)$,

$$Q_x = B\int_0^\infty J_0(\lambda\rho)e^{-l(z+a)}\frac{(m-l)\lambda}{(m+l)\,l}\,d\lambda,$$

$$R_x = -2B\int_0^\infty J_0(\lambda\rho)\,e^{mz-la}\frac{\lambda\, d\lambda}{l+m},$$

$$Q_z = A\int_0^\infty J_0(\lambda\rho)\,e^{-l(z+a)}\frac{h^2 l - k^2 m}{h^2 l + k^2 m}\frac{\lambda\, d\lambda}{l}$$

$$+ 2B\cos\phi\int_0^\infty J_1(\lambda\rho)\,e^{-l(z+a)}\frac{(h^2-k^2)\lambda^2 d\lambda}{(l+m)(h^2 l + k^2 m)},$$

$$R_z = 2Ah^2\int_0^\infty J_0(\lambda\rho)\,e^{mz-la}\frac{\lambda\, d\lambda}{h^2 l + k^2 m}$$

$$+ 2B\cos\phi\int_0^\infty J_1(\lambda\rho)\,e^{mz-la}\frac{(h^2-k^2)\lambda^2 d\lambda}{(l+m)(h^2 l + k^2 m)}.$$

The "directed effect" depends on the presence of the terms involving $\cos\phi$ in the expressions for Q_z and R_z. Now when $\sigma = \infty$ for the second medium, $h = \infty$, and these terms vanish altogether; hence the possibility of directing the energy of the radiation sent out from the bent antenna is due to the imperfect conductivity of the earth. Von Hoerschlemann has given a numerical discussion of the above formulae but the investigation is too long to be inserted here

§ 28. Propagation of electromagnetic waves along a straight wire of circular cross-section*.

Let us consider the symmetrical case when the electric force at any point is in a plane through the axis of the wire

* H. Hertz, *Electric Waves*; J. J. Thomson, *Proc. London Math. Soc.* Vol. 17 (1886), p. 310; *Recent Researches*, § 259; A. Sommerfeld, *Ann. d. Physik*, Bd. 67 (1899), p. 233; Gray and Mathews, *Bessel Functions*, Ch. 13; M. Abraham, *Encykl. d. Math. Wiss.* Band V. 2, Heft 3 (1910), p. 526; J. Larmor, *Proc. of the 5th Int. Congress of Mathematicians*, Vol. 1 (1912), p. 206. The problem considered in this last paper is chiefly that of alternating currents, viz. the *forced* alternations of flow produced by a uniform periodic electric force.

and the magnetic force is in circles at right angles to this plane. The field equations are then of the types

$$\left.\begin{aligned} \frac{\epsilon}{c}\frac{\partial E_z}{\partial t}+\frac{\sigma}{c}E_z=\frac{1}{\rho}\frac{\partial}{\partial\rho}(\rho H_\phi),\quad \frac{\epsilon}{c}\frac{\partial E_\rho}{\partial t}+\frac{\sigma}{c}E_\rho=-\frac{\partial H_\phi}{\partial z} \\ -\frac{\mu}{c}\frac{\partial H_\phi}{\partial t}=\frac{\partial E_\rho}{\partial z}-\frac{\partial E_z}{\partial \rho} \end{aligned}\right\} \quad (159).$$

These may be satisfied by putting

$$\left.\begin{aligned} E_z=-\frac{1}{\rho}\frac{\partial}{\partial\rho}\left(\rho\frac{\partial\Pi}{\partial\rho}\right),\qquad E_\rho=\frac{\partial^2\Pi}{\partial\rho\partial z} \\ H_\phi=-\frac{\epsilon}{c}\frac{\partial^2\Pi}{\partial t\partial\rho}-\frac{\sigma}{c}\frac{\partial\Pi}{\partial\rho} \end{aligned}\right\} \quad \text{.........(160)},$$

where Π satisfies the equation

$$\frac{\epsilon\mu}{c^2}\frac{\partial^2\Pi}{\partial t^2}+\frac{\sigma\mu}{c^2}\frac{\partial\Pi}{\partial t}=\frac{\partial^2\Pi}{\partial z^2}+\frac{1}{\rho}\frac{\partial}{\partial\rho}\left(\rho\frac{\partial\Pi}{\partial\rho}\right)=\Delta\Pi \quad \text{......(161)}.$$

Putting $\Pi=e^{i\omega t}u$, we find that $\Delta u+k^2u=0$, where

$$k^2=\frac{\epsilon\mu\omega^2-i\mu\omega\sigma}{c^2}.$$

We now assume that for points outside the wire*

$$u=A\,e^{-i\lambda z}\,K_0(\rho\sqrt{\lambda^2-k^2}),$$

and that for points inside the wire

$$u=Be^{-i\lambda z}\,J_0(i\rho\sqrt{\lambda^2-h^2}),$$

where h is the value of k inside the wire. These assumptions are made for the purpose of determining the periods and rates of decay of electric waves that can travel along the wire and maintain their own field. The first solution is chosen so as to make the flow of energy negligible at an infinite distance from the wire so that the system is self-contained; and to ensure this it is necessary to suppose that the real part of $\sqrt{\lambda^2-k^2}$ is positive. The second solution is chosen so as to make the electric and magnetic forces finite on the axis of the wire ($\rho=0$).

Let $\rho=a$ be the equation of the surface of the wire, then

* We follow here the work of Sommerfeld in which it is supposed that there are no conductors outside the wire. Sir J. J. Thomson allows for the presence of external conductors by supposing the dielectric surrounding the wire to be bounded by a cylindrical conductor having the same axis as the wire.

the continuity of the tangential components of the electric and magnetic forces requires that when $\rho = a$

$$\left.\begin{aligned} A\frac{\partial}{\partial\rho}\left[\rho K_0(\rho\sqrt{\lambda^2-k^2})\right] &= B\frac{\partial}{\partial\rho}\left[\rho J_0(i\rho\sqrt{\lambda^2-h^2})\right] \\ A\nu\frac{\partial}{\partial\rho}K_0(\rho\sqrt{\lambda^2-k^2}) &= B\nu_1\frac{\partial}{\partial\rho}J_0(i\rho\sqrt{\lambda^2-h^2}) \end{aligned}\right\} \quad (162),$$

ν and ν_1 being the values of the quantity $\sigma + i\omega\epsilon$ at points outside and inside the wire. The elimination of A and B gives rise to a transcendental equation for the determination of λ. The total current flowing along the wire is

$$\begin{aligned} J &= 2\pi\int_0^a \sigma E_z \rho\, d\rho \\ &= -2\pi\sigma e^{-i\lambda z}\rho B\frac{\partial}{\partial\rho}J_0(i\rho\sqrt{\lambda^2-h^2}) \quad \text{for } \rho = a. \end{aligned}$$

On the other hand the electric force $\bar{E}_z$ at the surface of the wire is

$$\begin{aligned} \bar{E}_z &= -\frac{1}{\rho}\frac{\partial}{\partial\rho}\left[\rho B e^{-i\lambda z}\frac{\partial}{\partial\rho}J_0(i\rho\sqrt{\lambda^2-h^2})\right] \quad \text{for } \rho = a \\ &= -B(\lambda^2-h^2)e^{-i\lambda z}J_0(ia\sqrt{\lambda^2-h^2}). \end{aligned}$$

Hence, if we put

$$\bar{E}_z = J\left\{R + \frac{i\omega}{c^2}L\right\},$$

where R and L are the resistance and self-induction of the wire, we get

$$R + \frac{i\omega}{c^2}L = \frac{\lambda^2-h^2}{2\pi\sigma}\left[\frac{J_0(x)}{xJ_0'(x)}\right]_{x=ia\sqrt{\lambda^2-h^2}}.$$

The roots of the transcendental equation have been discussed by Sommerfeld who used a method of successive approximations. The values of λ are complex as the waves which travel along the wire are damped owing to the imperfect conductance of the wire. It appears that when the disturbance does not penetrate far into the wire, the damping is small and so the velocity of propagation is very nearly equal to the velocity of light. When, however, the field does soak into the wire to some extent, the damping is of course considerable and so the wave travels with a velocity a little less than that of light. In the first case the real part of $\sqrt{\lambda^2-h^2}$ is large, in the second case it is small.

In Lecher's arrangement* there are two conjugate parallel wires between which the waves travel, consequently the field is not symmetrical round the axis of one of the wires. This case has been discussed by G. Mie† with the aid of bi-polar coordinates, and the lines of force have been studied by W. B. Morton‡. The latter also considers the case of n parallel wires passing through the corners of a regular polygon§.

The mathematical analysis for the case of a curved or twisted wire has not yet been fully developed. The important case of a spiral wire has, however, been discussed by H. C. Pocklington‖ and J. W. Nicholson¶. The latter gives numerous references to the literature of the subject. D. Hondros** has recently discussed the propagation of some types of unsymmetrical waves along a single wire. The electromagnetic theory of an electric cable has been given by Sir Joseph Thomson†† and F. Harms‡‡. The latter considers the case when the outer conductor of a cable is replaced by air.

§ 29. **Other problems which may be treated with cylindrical coordinates.**

The diffraction of electromagnetic waves by a cylindrical obstacle has been discussed by Lord Rayleigh§§, W. Seitz‖‖, W. Ignatowsky¶¶, P. Debye*†, C. Schaefer*‡, and J. W. Nicholson*§. Schaefer has made an extensive study of the case of a

* *Ann. d. Phys.* Bd. 41 (1890), p. 850. See also D. Mazzotto, *Il Nuovo Cimento* (4), t. 6 (1897), p. 172.

† *Ann. d. Phys.* Bd. 2 (1900), p. 201. The case in which the capacity is small is discussed at length by J. W. Nicholson, *Phil. Mag.* Feb.—Sept. (1909). The current is supposed to flow along one wire and return along the other.

‡ *Phil. Mag.* Vol. 50 (1900), p. 605; Vol. 4 (1902), p. 302.

§ *Phil. Mag.* Vol. 1 (1901), p. 563.

‖ *Proc. Camb. Phil. Soc.* Vol. 9 (1897), p. 324.

¶ *Phil. Mag.* Vol. 19 (1910), p. 77.

** *Ann. d. Phys.* Bd. 30 (1909), p. 905; *Dissertation*, Munich (1909).

†† *Proc. Roy. Soc.* Vol. 46 (1889), p. 1; *Recent Researches*, p. 262.

‡‡ *Ann. d. Phys.* Bd. 23 (1907), p. 44.

§§ *Phil. Mag.* Vol. 12 (1881), p. 81.

‖‖ *Ann. d. Phys.* Bd. 16 (1905), p. 746.

¶¶ *Ann. d. Phys.* Bd. 18 (1906), p. 495.

*† *Phys. Zeitschr.* (1908), p. 775.

*‡ *Ann. d. Phys.* Bd. 31 (1910), p. 462.

*§ *Proc. London Math. Soc.* Ser. 2, Vol. 11, p. 104.

dielectric cylinder and his results have been tested experimentally by Grossmann. This is one of the few cases in which the influence of the material properties of the obstacle has been taken into account in the mathematical treatment of a diffraction problem; the necessity of doing this has been clearly indicated by some of the experimental results.

P. Debye has studied the diffraction problem with reference to the theory of the rainbow. This had been done for the case of the sphere by L. Lorenz long ago*.

Electrical vibrations in regions bounded by cylinders of various shapes have been studied by Sir J. J. Thomson†, Lord Rayleigh‡, Sir J. Larmor§, R. H. Weber‖, A. Kalähne¶, and J. W. Nicholson**. The latter has also calculated the pressure exerted by a train of plane electromagnetic waves on a perfectly conducting cylinder††.

EXAMPLES.

1. An infinitely long metal cylinder of specific conductivity σ and permeability μ, bounded by the surface $r=R$, is surrounded by a dielectric of specific inductive capacity ϵ. A train of waves, in which the electric force is perpendicular to the cylinder and to magnetic force and if undisturbed would be represented by the real part of $Me^{i(\omega t+kr\cos\phi)}$, is passing in the dielectric. Prove that the magnetic force H_2 inside the cylinder and the part H_1 of the magnetic force outside representing the scattered wave, are given by the real parts of

$$H_1=e^{i\omega t}\sum_m a_m K_m(kr)\cos m\phi,\quad H_2=e^{i\omega t}\sum_m b_m J_m(hr)\cos m\phi,$$

where $h^2=-4\pi\mu\sigma i\omega$, and

$$b_m J_m(hR)-a_m K_m(kR)=2Mi^m J_m(kR),$$

$$\frac{b_m h}{4\pi\sigma}J_m'(hR)-\frac{a_m kc^2}{i\omega\epsilon}K_m'(kR)=\frac{2Mi^m kc^2}{i\omega\epsilon}J_m'(kR),$$

* *Oeuvres scientifiques*, pp. 405—502. See also Gans and Happel, *loc. cit.* (p. 44). The most recent paper on the rainbow is by W. Möbius, *Ann. d. Phys.* Bd. 33 (1910).

† *Recent Researches*, p. 344.

‡ *Phil. Mag.* (5), Vol. 43, p. 125.

§ *Proc. London Math. Soc.* (1), Vol. 26, p. 119.

‖ *Habilitationsschrift*, Heidelberg (1902); *Ann. d. Phys.* (4), Bd. 8 (1902), p. 721.

¶ *Ann. d. Phys.* (4), Bd. 19 (1906), pp. 80, 879; Bd. 18 (1905), p. 92.

** *Phil. Mag.* Aug. (1905), May (1906).

†† *Proc. London Math. Soc.* Ser. 2, Vol. 11, p. 104.

for positive integral values of m. The constants here have reference to the electromagnetic system of units.

(Cambr. Math. Tripos, Part II, 1905.)

2. Plane electromagnetic waves represented by

$$E_x = E_y = H_x = H_z = 0, \quad H_y = e^{ik(z+ct)}, \quad E_z = e^{ik(z+ct)},$$

fall upon the perfectly conducting cylinder $\rho = a$. Prove that in the scattered field

$$E_z = -\frac{1}{\pi} e^{ikct} \int_{-\infty}^{\infty} e^{ika \cos \beta} d\beta \int_{-\infty}^{\infty} \frac{K_\nu (ik\rho)}{K_\nu (ika)} \cos \nu\phi . \cos \nu\beta . d\nu.$$

(P. Debye.)

3. The wave-potential for a circular ring of point sources is given by

$$\Omega = \int_0^\infty e^{-z\sqrt{\lambda^2 - k^2} - ikct} J_0 (\lambda\rho) J_0 (\lambda a) \lambda d\lambda, \qquad z > 0.$$

(A. G. Webster.)

4. The wave-function

$$\Omega = e^{ikct} \int_0^\infty e^{-\lambda z} J_0 (\rho \sqrt{k^2 + \lambda^2}) \cos a\lambda \, d\lambda$$

is zero at points on the plane $z = 0$ which lie inside the circle $\rho^2 < a^2$; for $z = 0$, $\rho^2 > a^2$ its value is $e^{ikct} (\rho^2 - a^2)^{-\frac{1}{2}} \cos (k\sqrt{\rho^2 - a^2})$. (Sonin.)

5. If $R^2 = \rho^2 + (z + a \sinh u)^2$, $\xi = e^u$, the integral

$$\int_0^\infty J_0 (\lambda\rho) \lambda \, d\lambda \int_z^\infty J_0 \left[\lambda \sqrt{t^2 - z^2 + a^2 + a \left\{ \xi (t - z) + \frac{1}{\xi} (t + z) \right\}} \right] e^{ikt} dt$$

represents the function $\dfrac{e^{ikR}}{R} . e^{-a \cosh u}$ outside a paraboloid of revolution whose focus is at the singularity $z = -a \sinh u$, $\rho = 0$, and which passes through the circle $\rho = a$, $z = 0$. It is zero inside the paraboloid.

6. The circuital relations in cylindrical coordinates are

$$\frac{\rho}{c} \frac{\partial E_\rho}{\partial t} = \frac{\partial H_z}{\partial \phi} - \rho \frac{\partial H_\phi}{\partial z}, \qquad \frac{\rho}{c} \frac{\partial E_\phi}{\partial t} = \rho \frac{\partial H_\rho}{\partial z} - \rho \frac{\partial H_z}{\partial \rho},$$

$$\frac{\rho}{c} \frac{\partial E_z}{\partial t} = \frac{\partial}{\partial \rho} (\rho H_\phi) - \frac{\partial H_\rho}{\partial \phi}, \qquad \frac{\partial}{\partial \rho} (\rho H_\rho) + \frac{\partial H_\phi}{\partial \phi} + \rho \frac{\partial H_z}{\partial z} = 0$$

and similar equations in which E is replaced by $-H$ and H by E.

7. If $X = x \cos \omega t - y \sin \omega t$, $Y = x \sin \omega t + y \cos \omega t$, $Z = z - vt$, where v and ω are constants, the function $\Omega = F(X, Y, Z)$ is a wave-function if F satisfies the partial differential equation

$$\left(1 - \frac{\omega^2}{c^2} Y^2\right) \frac{\partial^2 F}{\partial X^2} + \left(1 - \frac{\omega^2}{c^2} X^2\right) \frac{\partial^2 F}{\partial Y^2} + 2 \frac{\omega^2}{c^2} XY \frac{\partial^2 F}{\partial X \partial Y} + \frac{\omega^2}{c^2} \left(X \frac{\partial F}{\partial X} + Y \frac{\partial F}{\partial Y} \right)$$

$$+ \left(1 - \frac{v^2}{c^2}\right) \frac{\partial^2 F}{\partial Z^2} = 0.$$

Obtain particular solutions of the form

$$\Omega = e^{\lambda(z-vt)+im(\phi-\omega t)}\left[AJ_m(p\rho)+BY_m(p\rho)\right]$$

where $p^2=\lambda^2\left(1-\frac{v^2}{c^2}\right)+\frac{m^2\omega^2}{c^2}$ and λ, m, A, B are arbitrary constants.

8. An oscillatory current is induced on a circular wire of radius a excited by a uniform electric force R_0e^{ipt} acting on its surface from the surrounding medium. Obtain expressions for the inductance and resistance of the wire per unit length when the wire is regarded as straight and no disturbing conductor is near.

CHAPTER V

THE PROBLEM OF DIFFRACTION

§ 30. Multiform solutions of the wave-equation*.

The wave-functions required to solve many of the boundary problems of Mathematical Physics are not single-valued functions of x, y, z, t in an ordinary space. We may, however, regard them as single-valued functions in a Riemann's space. This is a simple generalisation of the Riemann's surface of the theory of functions of a complex variable †; every plane section of the Riemann's space is in fact a Riemann's surface. Instead of branch lines and branch points we have branch membranes and branch curves. Thus in the physical problem of the diffraction of light through a circular hole in a screen, the boundary of the shadow of the screen is the branch membrane and the edge of the hole the branch curve.

We shall commence by finding a multiform solution of the equation

$$\frac{\partial^2 u}{\partial x^2} + \frac{\partial^2 u}{\partial y^2} + k^2 u = 0 \quad \ldots\ldots\ldots\ldots\ldots\ldots (163).$$

The fundamental solution $u = e^{ik(x \cos \alpha + y \sin \alpha)} = e^{ik\rho \cos(\phi - \alpha)}$ is of period 2π and can be expanded in the form

$$e^{ik\rho \cos(\phi - \alpha)} = J_0(k\rho) + 2 \sum_1^\infty i^n J_n(k\rho) \cos n(\phi - \alpha) \ldots (164).$$

* This theory is due to A. Sommerfeld, *Math. Ann.* Bd. 45 (1894), Bd. 47 (1896); *Zeitschr. für Math. u. Phys.* Bd. 46 (1901); *Proc. London Math. Soc.* (1), Vol. 28 (1897), p. 417. It has been developed by H. S. Carslaw, *Proc. London Math. Soc.* (1), Vol. 30, p. 121: (2), Vol. 8, p. 365; *Phil. Mag.* Vol. 5 (1903), p. 374: Vol. 20 (1910), p. 690; *Fourier's Series and Integrals*, Ch. 18; W. Voigt, *Gött. Nachr.* (1899); E. W. Hobson, *Camb. Phil. Trans.* Vol. 18 (1900), p. 277. Different methods have been used by H. M. Macdonald, *Electric Waves*, Appendix D; K. Schwarzschild, *Math. Ann.* Bd. 55 (1902), p. 177; *Proc. London Math. Soc.* Vol. 26 (1895), p. 156; C. W. Oseen, *Arkiv för mat.* Bd. 1 (1904), Bd. 2 (1905).

† See Harkness and Morley's *Theory of Functions* (1893), Ch. 6.

A solution of period $2m\pi$ may evidently be constructed by writing down a series of the form

$$\sum_{n=-\infty}^{\infty} a_n J_{\frac{n}{m}}(k\rho) \cos \frac{n}{m}(\phi - \alpha),$$

where the a_n's are suitable constants. The solution that seems the most natural extension of (164) is

$$F_m(\rho, \phi, \phi_0) = J_0(k\rho) + 2\sum_1^{\infty} i^{\frac{n}{m}} J_{\frac{n}{m}}(k\rho) \cos \frac{n}{m}(\phi - \phi_0) \quad \ldots(165).$$

To sum this series when $m = 2$, we transform the terms for which n is odd by means of the equation

$$J_{\frac{n}{2}}(k\rho) = \frac{(k\rho)^{\frac{n}{2}}}{2^{\frac{n}{2}}\Gamma(\frac{1}{2})\Gamma\left(\frac{n+1}{2}\right)} \int_0^{\pi} e^{ik\rho \cos \alpha} \sin^n \alpha \, . \, d\alpha.$$

Summing the two series separately we find that

$$F_2(\rho, \phi, \phi_0) = e^{ik\rho \cos(\phi - \phi_0)} + f(\phi - \phi_0) + f(\phi_0 - \phi),$$

where

$$f(\theta) = \left(\frac{ik\rho}{2\pi}\right)^{\frac{1}{2}} e^{\frac{i\theta}{2} + \frac{ik\rho}{2}e^{i\theta}} \int_0^{\pi} e^{-\frac{ik\rho}{2}e^{i\theta}\cos^2\alpha + ik\rho\cos\alpha} \sin \alpha \, . \, d\alpha$$

$$= \left(\frac{i}{\pi}\right)^{\frac{1}{2}} e^{ik\rho\cos\theta} \int_S^T e^{-i\lambda^2} d\lambda,$$

with $\quad T = \sqrt{2k\rho} \cos \dfrac{\theta}{2}, \quad S = \left(\dfrac{k\rho}{2}\right)^{\frac{1}{2}} \left(e^{-\frac{i\theta}{2}} - e^{\frac{i\theta}{2}}\right).$

Now
$$\int_{-S}^{S} e^{-i\lambda^2} d\lambda = 2\int_0^S e^{-i\lambda^2} d\lambda,$$

and
$$\int_{-\infty}^{0} e^{-i\lambda^2} d\lambda = \left(\frac{\pi}{i}\right)^{\frac{1}{2}};$$

hence we may write

$$F_2(\rho, \phi, \phi_0) = \left(\frac{i}{\pi}\right)^{\frac{1}{2}} e^{ik\rho\cos(\phi - \phi_0)} \int_{-\infty}^{\tau} e^{-i\lambda^2} d\lambda \quad \ldots(166),$$

where $\quad \tau = \sqrt{2k\rho} \cos \frac{1}{2}(\phi - \phi_0).$

With the aid of the function F_2 we can solve some problems on the diffraction of plane electromagnetic waves by a semi-infinite plane bounded by a straight edge. Let us consider the case of a totally reflecting screen*. If the electric force in the incident wave is parallel to the edge of the screen, the electric force $u = E_z$ for the total disturbance must vanish over both faces of the screen and must satisfy the differential equation (163). These conditions are fulfilled by taking

$$E_z = F_2(\rho, \phi, \phi_0) - F_2(\rho, \phi, -\phi_0) \quad \text{.........(167)}.$$

This value of E_z also satisfies the right conditions at infinity. To prove this we must find an asymptotic expression for F_2 when ρ is large.

Now when $\tau > 0$, we have the asymptotic expansion †

$$\int_\tau^\infty e^{-i\lambda^2} d\lambda \sim -\frac{ie^{-i\tau^2}}{\tau}\left[1 - \frac{1}{2i\tau^2} + \frac{1\cdot 3}{(2i\tau^2)^2} - \ldots\right];$$

while when $\tau < 0$, $\int_{-\infty}^{\tau}$ has a similar asymptotic expansion with the sign changed. This means that when $\cos \frac{1}{2}(\phi - \phi_0) > \epsilon > 0$ we have

$$F_2(\rho, \phi, \phi_0) \sim e^{ik\rho\cos(\phi - \phi_0)} + \frac{i^{\frac{3}{2}}}{\sqrt{2\pi k\rho}\cos\frac{1}{2}(\phi - \phi_0)} e^{-ik\rho} - \ldots,$$

while when $\cos \frac{1}{2}(\phi - \phi_0) < \epsilon < 0$ there is a similar asymptotic expansion in which the first term is missing. It thus appears that the electric force in the geometric shadow vanishes at infinity to the order $\rho^{-\frac{1}{2}}$.

If the magnetic force in the incident wave is parallel to the edge of the screen, the magnetic force $u = H_z$ must satisfy the differential equation (163) and be such that $\frac{\partial u}{\partial \phi} = 0$ over both faces of the screen. The conditions are fulfilled by putting

$$H_z = F_2(\rho, \phi, \phi_0) + F_2(\rho, \phi, -\phi_0) \quad \text{.........(168)}.$$

Prof. H. M. Macdonald has shown that the solution of a problem concerning a perfectly absorbing body can be made to depend on the solution of two allied problems‡. "A per-

* The incident waves are supposed to come in a direction for which $\phi = \pi + \phi_0$. An approximate solution of this problem was given by H. Poincaré, *Acta Mathematica*, Bd. 16, p. 297; Bd. 20, p. 313.

† Bromwich's *Infinite Series*, p. 328.

‡ *Phil. Trans.* A, Vol. 212 (1912), p. 337; *Proc. London Math. Soc.* (2), Vol. 12 (1913).

fectly absorbing body may be regarded as a body which is incapable of supporting either electric or magnetic force; hence if C is the electric current distribution on the surface of the body when it is supposed to be perfectly conducting, and C' is the magnetic current distribution on the surface of the body when it is supposed to be incapable of supporting magnetic force, the superposition of these two distributions gives the electric and magnetic current distributions on the surface of the body when it is perfectly absorbing and the amplitude of the incident waves is doubled."

Now if we suppose our screen to be incapable of supporting magnetic force, the boundary condition is that the tangential component of the magnetic force should vanish. When the electric force is parallel to the axis of z, $\frac{\partial E_z}{\partial \phi}$ must vanish over the screen. Hence

$$E_z = F_2(\rho, \phi, \phi_0) + F_2(\rho, \phi, -\phi_0) \quad \text{.........(169)}.$$

The solution for a totally absorbing screen is thus simply *

$$E_z = F_2(\rho, \phi, \phi_0) \quad \text{.................(170)}.$$

Similarly, it can be shown that when the magnetic force is parallel to the axis of z, the solution for a perfectly absorbing screen is

$$H_z = F_2(\rho, \phi, \phi_0) \quad \text{..................(171)}.$$

Prof. Lamb † has discussed the case of perpendicular incidence with the aid of the parabolic substitution

$$\left.\begin{array}{lll} \xi = \rho^{\frac{1}{2}} \cos \frac{1}{2}\phi, & \eta = \rho^{\frac{1}{2}} \sin \frac{1}{2}\phi & \\ x = \xi^2 - \eta^2, & y = 2\xi\eta, & \rho = \xi^2 + \eta^2 \end{array}\right\} \text{...(172)}.$$

The curves $\xi = \text{const.}$, $\eta = \text{const.}$ are confocal parabolas, $\eta = 0$ is the screen. A solution of Maxwell's equations is obtained by writing

$$\left.\begin{array}{lll} H_x = 0, & H_y = 0, & H_z = u \\ \frac{\partial E_x}{\partial t} = c\frac{\partial u}{\partial y}, & \frac{\partial E_y}{\partial t} = -c\frac{\partial u}{\partial x}, & \frac{\partial E_z}{\partial t} = 0 \end{array}\right\} \text{...(173)},$$

where

$$\frac{\partial^2 u}{\partial x^2} + \frac{\partial^2 u}{\partial y^2} = \frac{1}{c^2}\frac{\partial^2 u}{\partial t^2} \quad \text{..............(174)}.$$

* W. Voigt, *Gött. Nachr.* (1899), p. 1, discusses the case of an absorbing screen.

† *Proc. London Math. Soc.* (2), Vol. 4 (1907), p. 190; Vol. 8 (1910), p. 422.

Starting with Poisson's wave-function* $\chi = \rho^{-\frac{1}{2}} \cos \frac{1}{2}\phi \,.\, f(ct - \rho)$, let us put

$$\frac{\partial u}{\partial x} = \chi.$$

Transforming to the coordinates ξ, η, we obtain

$$\xi \frac{\partial u}{\partial \xi} - \eta \frac{\partial u}{\partial \eta} = 2\xi f(ct - \xi^2 - \eta^2) \quad \text{.........(175).}$$

Solving this partial differential equation by Lagrange's method and adjusting the complementary function so that the boundary condition $\frac{\partial u}{\partial \eta} = 0$ for $\eta = 0$ is satisfied, we obtain

$$u = \int_0^{\xi+\eta} f(ct + y - \zeta^2)\, d\zeta + \int_0^{\xi-\eta} f(ct - y - \zeta^2)\, d\zeta$$
$$+ \tfrac{1}{2} F(ct + y) + \tfrac{1}{2} F(ct - y) \text{...(176),}$$

where F is an arbitrary function. If the boundary condition is $u = 0$ for $\eta = 0$ the sign of the second term must be changed. It is easy to verify that each of the integrals represents a wave-function.

Let us now put $\quad f(x) = \frac{2}{\pi}\int_0^\infty F'(x - v^2)\, dv,$

and make the substitution $\zeta = \sigma \cos \alpha$, $v = \sigma \sin \alpha$, then after a little reduction we find that†

$$\left.\begin{aligned}
u &= F(ct + y) - \frac{1}{\pi}\int_0^{\frac{\pi}{2}} F[ct + y - (\rho + y)\sec^2 \alpha]\, d\alpha \\
&\quad + \frac{1}{\pi}\int_0^{\frac{\pi}{2}} F[ct - y - (\rho - y)\sec^2 \alpha]\, d\alpha \qquad x < 0,\ y \gtrless 0 \\
u &= F(ct + y) + F(ct - y) - \frac{1}{\pi}\int_0^{\frac{\pi}{2}} F[ct + y - (\rho + y)\sec^2 \alpha]\, d\alpha \\
&\quad - \frac{1}{\pi}\int_0^{\frac{\pi}{2}} F[ct - y - (\rho - y)\sec^2 \alpha]\, d\alpha \qquad \begin{matrix} x > 0 \\ y > 0 \end{matrix} \\
u &= \frac{1}{\pi}\int_0^{\frac{\pi}{2}} F[ct + y - (\rho + y)\sec^2 \alpha]\, d\alpha \\
&\quad + \frac{1}{\pi}\int_0^{\frac{\pi}{2}} F[ct - y - (\rho - y)\sec^2 \alpha]\, d\alpha \qquad \begin{matrix} x > 0 \\ y > 0 \end{matrix}
\end{aligned}\right\} \text{.........(177).}$$

* *Journal de l'École Polytechnique*, Cah. 19, t. 12 (1823). See also V. Volterra, *Acta Math.* t. 18; Häntzschel, *Reduction der Potentialgleichung*, Ch. 1.

† These formulae are not given in Prof. Lamb's paper.

Each of the integrals represents a wave-function provided $F(\pm\infty)=0$ and

$$\tan\alpha\, F'[ct \pm y - (\rho \pm y)\sec^2\alpha] \to 0 \text{ as } \alpha \to \frac{\pi}{2}.$$

In these circumstances we have the solution of the diffraction problem for the case when the initial disturbance is represented by $u=F(ct+y)$, the magnetic force being parallel to the axis of z. By suitably choosing F we can deal with the case of a solitary wave.

A new method of solving the problem of diffraction by a straight edge has been given recently by Oseen*.

Problems connected with a wedge have been treated successfully by Sommerfeld and other writers by using a certain type of contour integral. The fundamental solution of (163) is now

$$u=\frac{1}{2\pi i}\int e^{ik\rho\cos(\phi-\alpha)}\frac{dv}{d\alpha}\cdot\frac{d\alpha}{v} \quad \text{............(178)},$$

where $v=e^{\frac{i\alpha}{n}}-e^{\frac{i\phi_0}{n}}$ and the path of integration is a simple contour which starts from $\infty i+\gamma$ and goes to $\infty i+\gamma'$ without crossing the real axis. The quantities γ, γ' are subject to the inequalities

$$2\pi>\gamma>\pi, \qquad 0>\gamma'>-\pi.$$

This function u is multiform and of period $2n\pi$, but on an n-sheeted Riemann's surface with the origin as branch-point and the line $\phi=-(\pi-\phi_0)$ as branch-section, it is uniform.

With the aid of this function a number of diffraction problems may be solved.

Thus in the case of a perfectly conducting prism of angle $2\pi-\alpha$ if the electric force in the incident waves is parallel to the edge of the screen and is represented by the real part of the expression

$$e^{ik[ct+\rho\cos(\phi-\phi_0)]},$$

it can be shown by an extension of the method of images, that for the total disturbance the electric force is the real part of the expression†

$$E_z=\frac{1}{2\pi i}e^{ikct}\int e^{ik\rho\cos\zeta}\frac{d}{d\zeta}\log\frac{w}{w_1}\,d\zeta \quad \text{......(179)},$$

* *Arkiv för matematik, astronomi och fysik* (1912).

† Macdonald, *Electric Waves*, p. 192 (1902).

where

$$w = \cos\frac{\pi\zeta}{\alpha} - \cos\frac{\pi}{\alpha}(\phi - \phi_0), \qquad w_1 = \cos\frac{\pi\zeta}{\alpha} - \cos\frac{\pi}{\alpha}(\phi + \phi_0),$$

and the path of integration is the same as before.

In the associated problem when the magnetic force in the incident wave is parallel to the axis of z, the magnetic force for the total disturbance is the real part of the expression

$$H_z = \frac{1}{2\pi i} e^{ikct} \int e^{ik\rho\cos\zeta} \frac{d}{d\zeta} \log (ww_1)\, d\zeta \quad \text{......(180)}.$$

These solutions and the solutions of analogous problems have been discussed by W. H. Jackson *, H. M. Macdonald †, F. Reiche ‡, A. Wiegrefe §, and other writers.

§ 31. **Elliptic coordinates** ‖.

If we put

$$x = \cosh\omega\cos\chi, \qquad y = \sinh\omega\sin\chi \quad \text{......(181)},$$

the differential equation (163) becomes

$$\frac{\partial^2 u}{\partial\omega^2} + \frac{\partial^2 u}{\partial\chi^2} + k^2(\cosh^2\omega - \cos^2\chi)\, u = 0 \quad \text{......(182)}.$$

The elementary solutions are now of the form

$$u = E(\omega)\, F(\chi),$$

where E and F satisfy the equations of the elliptic cylinder ¶

$$\left.\begin{aligned} \frac{d^2E}{d\omega^2} + (k^2\cosh^2\omega + p)\, E = 0 \\ \frac{d^2F}{d\chi^2} - (k^2\cos^2\chi + p)\, F = 0 \end{aligned}\right\} \quad \text{............(183)}.$$

Appropriate solutions of these differential equations have been obtained recently by Prof. Whittaker **.

* *Proc. London Math. Soc.* Ser. 2, Vol. 1 (1904), p. 393.

† *Ibid.* Vol. 12 (1913), p. 430.

‡ *Ann. d. Phys.* Bd. 37 (1912), p. 131.

§ *Ibid.* Vol. 39 (1912), p. 449.

‖ H. Weber, *Math. Ann.* Vol. 1 (1869); Mathieu, *Louville's Journal*, Ser. 2, Vol. 13 (1868); Hartenstein, *Hoppe's Archiv* (2), t. 14, p. 170; R. C. Maclaurin, *Cambr. Phil. Trans.* Vol. 17 (1898), p. 41.

¶ For this equation see Heine, *Handbuch der Kugelfunktionen*; Lindemann, *Math. Ann.* Bd. 22; Häntzschel, *Zeitschr. f. Math. u. Phys.* Vol. 31, p. 25 (1883); Mathieu, *Liouville's Journal*, Ser. 2, Vol. 13 (1868).

** *Math. Congress*, Cambridge (1912).

Elliptic coordinates are appropriate for the solution of problems connected with the scattering of electromagnetic waves by an elliptic cylinder*.

W. Wien has suggested † that the problem of the diffraction of light through a straight slit in a screen ‡ may be treated with the aid of elliptic coordinates by regarding the screen as a limiting case of a hyperbolic cylinder.

H. Weber § has shown that when $k \neq 0$ the elliptic and parabolic substitutions are the only transformations which lead to elementary solutions of the equation (163). For further properties of this differential equation we may refer to Pockels, *Über die partielle Differentialgleichung* $\Delta u + k^2 u = 0$, Teubner, Leipzig (1891), and to Lord Rayleigh's *Theory of Sound*.

§ 32. **Other diffraction problems.**

The diffraction of light and electric waves by a grating of wires is a problem of importance, but the mathematical treatment is very difficult and the theories that have been given so far are of an approximate character. Sir J. J. Thomson|| has discussed the theory of Hertz's grating ¶ which consists of a number of parallel equidistant metal wires. When electric waves whose wave-length is large compared with the distance between the wires fall normally on the grating, they pass through if the electric force is at right angles to the wires but are reflected if the electric force is parallel to the wires. Prof. Lamb ** has considered the case of a grating which consists of parallel strips of metal; his theory has been supported by the

* See, for instance, K. Aichi, *Proc. Tokyo Math. Phys. Soc.* (2), 4, p. 266 (1908); B. Sieger, *Ann. d. Physik* (4), Bd. 27 (1908), p. 626.

† *Jahresbericht d. deutsch. Math. Verein*, Bd. 15 (1906), p. 42.

‡ For this problem see K. Schwarzschild, *loc. cit.*; Lord Rayleigh, *Phil. Mag.* Vol. 43 (1897), p. 259; *Scientific Papers*, Vol. 4, p. 283; *Proc. Roy. Soc.* A, Vol. 89 (1913), p. 194. An interesting experimental result has been obtained recently by P. Zeeman, *Amsterdam Proc.*, Nov. 28 (1912), p. 599.

§ *Math. Ann.* Bd. 1. See also Häntzschel, *Reduction der Potentialgleichung*, p. 137.

|| *Recent Researches* (1893), p. 425.

¶ *Collected Works*, Vol. 2, p. 190. For some recent experimental work see H. du Bois and H. Rubens, *Ann. d. Phys.* Bd. 35 (1911), p. 243; A. D. Cole, *Phys. Review*, Jan. (1913).

** *Proc. London Math. Soc.* Ser. 1, Vol. 29 (1898), p. 523.

experimental work of Cl. Schaefer*, J. Langwitz*, and G. H. Thomson†.

Lord Rayleigh‡ has given an approximate electromagnetic theory of the action of a grating on waves of light and this theory has been extended by W. Voigt§ so as to take into account the properties of the material of which the grating is made. Voigt's theory has been tested experimentally by B. Pogany|| who gives an account of previous experimental work on the subject.

The diffraction of light through a circular hole in a screen is a problem of interest to mathematicians which has yet to be solved¶. A promising method of attack is to regard the screen as the limiting case of a hyperboloid of revolution of one sheet.

Examples. 1. Prove that

$$e^{ikct}F_2(\rho, \phi, \phi_0) = \left(\frac{ikc}{\pi}\right)^{\frac{1}{2}} \int_{-\infty}^{t-\frac{\rho}{c}} \frac{e^{ikc\tau}d\tau}{\sqrt{c(t-\tau)+\rho\cos(\phi-\phi_0)}}.$$

2. If $x+iy = a\cosh(\omega+i\chi)$, $\xi = \frac{1}{2}kae^{\omega}$, $\eta = \frac{1}{2}kae^{-\omega}$,

$$J_\nu(k\rho)e^{\pm i\nu\phi} = \sum_n J_{\frac{\nu+n}{2}}(\xi) J_{\frac{\nu-n}{2}}(\eta) e^{\pm in\chi},$$

where the summation extends over all even integral values of n if ν is even and over all odd integral values of n if ν is odd.

(J. H. Hartenstein, *Grunert's Archiv* (2), t. 14, p. 170.)

§ 32*a*. The introduction and elimination of discontinuities.

Wave-functions with singular lines or with singularities travelling along straight lines with the velocity of light may sometimes be employed with advantage in the solution of diffraction problems. To illustrate the method to be adopted we shall consider the diffraction of waves of sound by an

* *Ann. d. Phys.* Bd. 21 (1906), p. 587. The theory is developed from a new point of view by C. Schaefer and F. Reiche, *Ibid.* Bd. 32 (1910), p. 577; Bd. 35 (1911), p. 817.

† *Ibid.* Vol. 22 (1907), p. 365.

‡ *Proc. Roy. Soc.* A, Vol. 79 (1907), p. 532.

§ *Gött. Nachr.* (1911).

|| *Ann. d. Phys.* Bd. 37 (1912), p. 257.

¶ Approximate solutions have been given by G. G. Stokes, *Camb. Phil. Trans.* (1849); H. Lorenz, *Vidensk. Selsk. Skr.*, Copenhagen (1890); H. A. Rowland, *Amer. Journ.* Vol. 6; A. Grimpen, *Diss.* Kiel (1890); A. E. H. Love, *Phil. Trans.* A, Vol. 197 (1901).

infinitely thin semi-infinite plane bounded by a straight edge*. We shall suppose that the waves are sent out from a stationary source and that the screen acts as a perfect reflector†.

Let the axis of z be taken along the edge of the screen and the axis of x in the plane of the screen at right angles to the edge. Let P be the source of sound and Q an arbitrary point on the edge of the screen. If the disturbance issuing from P were reflected according to the laws of geometrical optics, the total disturbance would be discontinuous in crossing two semi-infinite planes, each of which is bounded by the edge of the screen. The first of these planes is a boundary of the geometrical shadow, when continued across the edge of the screen it passes through P. The second plane is the boundary of the geometrical shadow for the optical image of P, viz. P_1.

To obtain the correct solution of the diffraction problem we must add to the disturbance just described a second one having discontinuities which will annul the above-mentioned discontinuities, the new disturbance must also be chosen so that the boundary condition is satisfied at the two faces of the screen. We shall now show that the required disturbance can be built up by superposition from elementary disturbances with singularities along lines such as PQ and P_1Q produced.

Let R be the distance of an arbitrary point (x, y, z, t) from Q, then if $(0, 0, \zeta)$ are the coordinates of Q and c is the velocity of sound, we know that a function of type

$$\frac{1}{R} f\left(t - \frac{R}{c}, \ \frac{x + iy}{z - \zeta + R}\right)$$

satisfies the wave-equation. Let us choose the arbitrary function f in such a way that the expression becomes infinite along the line PQ produced and returns to its initial value when the point x, y, z is rotated twice round the edge of the screen. This last condition is added so as to enable us to satisfy the boundary condition.

* This problem has been solved by H. S. Carslaw, *Proc. London Math. Soc.* Vol. 30 (1898), p. 121. A transformation of his solution suggested the method described here.

† This assumption is usually justifiable. In Prof. A. G. Webster's experiments on the reflection of sound from the ground (*Phys. Review*, Vol. 28 (1909), p. 65) it was found that the reflection is more than 90 %.

Now if we write

$$\frac{x+iy}{z-\zeta+R}=e^{-u+i\phi},$$

where u and ϕ are real quantities, and use R_0, u_0, ϕ_0 to denote the values of R, u, ϕ for the point P, it is easy to see that the function

$$\frac{1}{R}F\left(t-\frac{R}{c}\right)\sec\tfrac{1}{2}\left[\phi-\phi_0+i\left(u+u_0\right)\right]$$

satisfies the requirements, for it is periodic in ϕ with period 4π and is infinite along PQ produced, where

$$u=-u_0, \quad \phi=\phi_0+\pi$$

We now imagine sources corresponding to wave-functions of this type to be associated with each element $d\zeta$ of the edge, and suppose the strength and phase of the source at Q to depend on its position relative to P in such a way that

$$F\left(t-\frac{R}{c}\right)=\frac{1}{R_0}f\left(t-\frac{R+R_0}{c}\right),$$

where $f(t)$ is the strength of the source at P at time t.

In this way we obtain an integral

$$V_1=\frac{1}{4\pi}\int_{-\infty}^{\infty}\frac{d\zeta}{RR_0}f\left(t-\frac{R+R_0}{c}\right)\sec\tfrac{1}{2}\left[\phi-\phi_0+i\left(u+u_0\right)\right],$$

which will be shown to be discontinuous in the way required as the point (x, y, z) crosses the boundary of the shadow for P.

In a similar way we can construct an integral

$$V_2=\frac{1}{4\pi}\int_{-\infty}^{\infty}\frac{d\zeta}{RR_0}f\left(t-\frac{R+R_0}{c}\right)\sec\tfrac{1}{2}\left[\phi+\phi_0-i\left(u+u_0\right)\right],$$

which can be shown to be discontinuous in the way required as the point (x, y, z) crosses the boundary of the geometrical shadow for P_1.

Now let r, r_1 be the distances of the point x, y, z respectively from P and P_1, then the velocity potential V of the total disturbance is given by the following expressions in the different regions of space

$$V=\frac{1}{r}f\left(t-\frac{r}{c}\right)+\frac{1}{r_1}f\left(t-\frac{r_1}{c}\right)-V_1-V_2 \quad \text{in } S_1,$$

$$V=\frac{1}{r}f\left(t-\frac{r}{c}\right)-V_1-V_2 \quad \text{in } S_2,$$

$$V=-V_1-V_2 \quad \text{in } S_3.$$

The space S_1 is bounded by the screen and the limiting plane of the geometrical shadow for P_1, S_2 is bounded by the limiting planes of the shadows for P and P_1, S_3 is bounded by the screen and the limiting plane of the shadow for P.

The boundary condition is that $\frac{\partial V}{\partial \phi}$ should be zero over the two faces of the screen and it is easy to verify that this condition is satisfied. To show that V is continuous for the whole of the space outside the screen and vanishes at infinity when the function f is finite, we shall transform the integrals V_1, V_2 to the forms given by Prof. Carslaw. To do this we put $u + u_0 = b$, then if ρ is the distance of a point from the axis of z, we have

$$\rho e^{u} = z - \zeta + R, \qquad \rho_0 e^{u_0} = z_0 - \zeta + R_0,$$
$$R^2 = \rho^2 + (z - \zeta)^2, \qquad R_0^2 = \rho_0^2 + (z_0 - \zeta)^2,$$
$$\rho e^{-u} = R - z + \zeta, \qquad \rho_0 e^{-u_0} = R_0 - z_0 + \zeta,$$
$$\rho\rho_0 \cosh(u + u_0) = (z - \zeta)(z_0 - \zeta) + RR_0,$$
$$\rho^2 + \rho_0^2 + (z - z_0)^2 + 2\rho\rho_0 \cosh(u + u_0) = (R + R_0)^2,$$
$$db = du + du_0 = -d\zeta\left(\frac{1}{R} + \frac{1}{R_0}\right).$$

Hence it follows that

$$V_1 = \frac{1}{4\pi}\int_{-\infty}^{\infty} \frac{db}{R + R_0} f\left(t - \frac{R + R_0}{c}\right) \sec \tfrac{1}{2}(\phi - \phi_0 + ib).$$

On substituting the expression for $R + R_0$ in terms of b we obtain an integral which is equivalent to the one given by Prof. Carslaw. To see that it is discontinuous* we use V_+ and V_- to denote the values of the integral for $\phi = \pi + \phi_0 + \epsilon$ and $\phi = \pi + \phi_0 - \epsilon$ respectively, where ϵ is a small quantity. The difference between these quantities may be regarded as a contour integral and can be evaluated by Cauchy's theorem We may write

$$V_+ - V_- = -\frac{1}{4\pi i}\int \frac{d\xi}{R + R_0} f\left(t - \frac{R + R_0}{c}\right) \operatorname{cosec} \tfrac{1}{2}\xi.$$

Taking the residue for $\xi = 0$, i.e. $\phi = \pi + \phi_0$, $b = 0$, we get

$$V_+ - V_- = -\frac{1}{R + R_0} f\left(t - \frac{R + R_0}{c}\right) = -\frac{1}{r} f\left(t - \frac{r}{c}\right);$$

* A more careful proof is given in Prof. Carslaw's paper.

for the conditions $\phi = \pi + \phi_0$, $b = 0$ imply that the radii R and R_0 are in one straight line and so give r when added together.

It is now clear that the integral V_1 possesses the right type of discontinuity and a similar remark holds for the integral V_2. The method can no doubt be modified so as to give solutions of other types of diffraction problems, the chief difficulty arises in the choice of a function which will satisfy the boundary conditions. At any rate the method suggests an interesting type of boundary problem in which the desired wave-functions have specified discontinuities instead of being continuous everywhere. This type of problem ought to be studied more completely.

In the general problem of the diffraction round a moving object of the waves issuing from a moving source, the wave-functions that are derived by the methods of geometrical optics have discontinuities at a certain boundary which is the locus of points travelling along straight lines with the velocity of light.

The points in question start from certain points of the moving object and move along tangents to the surface of the object, their paths being in fact continuations of the paths of particles that may be considered to have been emitted from the source. Indeed, if we imagine the source to emit particles in all directions as it moves about, the particles which just graze the moving object will, when they continue their rectilinear motion with the velocity of light, form the boundary at which the discontinuities arise.

In Chapter VIII we shall obtain a class of wave-functions with singularities moving along straight lines with the velocity of light. These functions seem to be just the ones that are required for the building up of wave-functions with discontinuities of the type just described. The problem of forming in this way the functions which will enable us to complete the solution of the diffraction problem is one which awaits solution.

CHAPTER VI

TRANSFORMATIONS OF COORDINATES APPROPRIATE FOR THE TREATMENT OF PROBLEMS CONNECTED WITH A SURFACE OF REVOLUTION

§ 33. Spheroidal coordinates.

Problems in which there is symmetry round the axis of z can often be treated with the aid of a substitution of the form*

$$\rho + iz = f(\alpha + i\beta) \quad \text{.................(184).}$$

Taking α, β, ϕ as orthogonal coordinates, we have

$$dx^2 + dy^2 + dz^2 = (d\alpha^2 + d\beta^2)\frac{\partial(\rho, z)}{\partial(\alpha, \beta)} + \rho^2 d\phi^2 \text{...(185),}$$

and equations (18) of § 8 become

$$\left.\begin{aligned}
\frac{J}{\rho}\left[\frac{\partial}{\partial\beta}(\rho M_\phi) - \frac{\partial}{\partial\phi}\left(\frac{1}{J}M_\beta\right)\right] &= \pm kM_\alpha \\
\frac{J}{\rho}\left[\frac{\partial}{\partial\phi}\left(\frac{1}{J}M_\alpha\right) - \frac{\partial}{\partial\alpha}(\rho M_\phi)\right] &= \pm kM_\beta \\
J^2\left[\frac{\partial}{\partial\alpha}\left(\frac{1}{J}M_\beta\right) - \frac{\partial}{\partial\beta}\left(\frac{1}{J}M_\alpha\right)\right] &= \pm kM_\phi
\end{aligned}\right\} \text{......(186),}$$

$$\frac{\partial}{\partial\alpha}\left(\frac{\rho}{J}M_\alpha\right) + \frac{\partial}{\partial\beta}\left(\frac{\rho}{J}M_\beta\right) + \frac{\partial}{\partial\phi}\left(\frac{1}{J^2}M_\phi\right) = 0,$$

where
$$J^2 = \frac{\partial(\alpha, \beta)}{\partial(\rho, z)}.$$

* This substitution has been used in other branches of mathematical physics by C. Neumann, *Theorie der Elektricitäts- und Wärme-Vertheilung in einem Ringe* (1864); E. Mathieu, *Cours de physique mathématique* (1873); A. Wangerin, *Berliner Monatsberichte* (1878); Häntzschel, *Reduction der Potentialgleichung*; Michell, *Mess. of Math.* (1890); Basset, *Hydrodynamics*, Vol. 2, p. 8; F. H. Safford, *Amer. Journ.* Vol. 21; *Archiv der Math.* Bd. 13 (1908), p. 22. The important developments on which the following analysis is founded are contained in papers to which we shall refer presently.

These equations may be satisfied by putting

$$M_\alpha = \frac{J}{\rho}\frac{\partial \Omega}{\partial \beta}, \quad M_\beta = -\frac{J}{\rho}\frac{\partial \Omega}{\partial \alpha}, \quad M_\phi = \pm \frac{k}{\rho}\Omega \quad \text{...(187)},$$

where $\Omega = U \pm iV$ is a solution of the partial differential equation

$$\frac{\partial^2 \Omega}{\partial \alpha^2} + \frac{\partial^2 \Omega}{\partial \beta^2} - \frac{1}{\rho}\left(\frac{\partial \rho}{\partial \alpha}\frac{\partial \Omega}{\partial \alpha} + \frac{\partial \rho}{\partial \beta}\frac{\partial \Omega}{\partial \beta}\right) + \frac{k^2}{J^2}\Omega = 0 \quad \text{...(188)}.$$

The problem of finding the periods of free electrical oscillations on a conducting spheroid is of considerable interest because a straight rod of circular cross-section can be regarded as approximately equivalent to a prolate spheroid whose major axis is relatively much longer than the minor axis. This problem has been treated very fully by M. Abraham*, R. C. Maclaurin†, M. Brillouin‡, F. Ehrenhaft§ and J. W. Nicholson‖. The effect of a spheroidal obstacle on a train of waves has been studied by K. F. Herzfeld¶.

For prolate spheroids the appropriate substitution is**

$$z + i\rho = a\cosh(\alpha + i\beta) \quad \text{..................(189)},$$

or

$$\rho = a\sinh\alpha\sin\beta, \quad z = a\cosh\alpha\cos\beta,$$

$$\frac{\partial(\rho, z)}{\partial(\alpha, \beta)} = -a^2(\cosh^2\alpha - \cos^2\beta) \quad \text{.........(190)}.$$

The partial differential equation is now

$$\frac{\partial^2 \Omega}{\partial \alpha^2} + \frac{\partial^2 \Omega}{\partial \beta^2} - \coth\alpha\frac{\partial \Omega}{\partial \alpha} - \cot\beta\frac{\partial \Omega}{\partial \beta} - a^2k^2(\cosh^2\alpha - \cos^2\beta)\,\Omega \quad \text{......(191)},$$

and there are elementary solutions of the form $\Omega = A(\alpha)B(\beta)$ where A and B satisfy the differential equations

$$\left.\begin{aligned} \frac{d^2A}{d\alpha^2} - \coth\alpha\frac{dA}{d\alpha} - (k^2a^2\cosh^2\alpha + \lambda)A = 0 \\ \frac{d^2B}{d\beta^2} - \cot\beta\frac{dB}{d\beta} + (\lambda + k^2a^2\cos^2\beta)B = 0 \end{aligned}\right\} \quad \text{...(192)}.$$

* *Dissertation*, Berlin (1897); *Ann. d. Phys.* Bd. 66 (1898), p. 435; *Math. Ann.* Bd. 52 (1899), p. 81.

† *Cambr. Phil. Trans.* Vol. 17 (1898–9), pp. 41—108.

‡ *Propagation de l'électricité* (1904), Ch. VI.

§ *Wiener Berichte* (1904), p. 273.

‖ *Phil. Mag.* (1906).

¶ *Wiener Berichte* (1911), p. 1587.

** Cf. Heine, *Crelle*, Bd. 26 (1843), p. 185; *Kugelfunktionen*, Bd. 2, § 38; Lamb's *Hydrodynamics*, p. 132.

These equations are discussed in some detail in the papers to which we have just referred, they may be reduced to a special form of an equation obtained by Prof. C. Niven* in a study of the conduction of heat in ellipsoids of revolution.

For oblate spheroids the appropriate substitution is

$$\rho + iz = a \cosh (\alpha + i\beta) \quad \text{...............(193)},$$

giving $\quad \rho = a \cosh \alpha \cos \beta, \quad z = a \sinh \alpha \sin \beta,$

$$\frac{\partial (\rho, z)}{\partial (\alpha, \beta)} = a^2 (\cosh^2 \alpha - \cos^2 \beta).$$

The surfaces $\beta = \text{const.}$ are now hyperboloids of one sheet, the surface $\beta = 0$ can be regarded as the surface of a screen which is pierced by a circular hole of radius a.

The partial differential equation for Ω is now

$$\frac{\partial^2 \Omega}{\partial \alpha^2} + \frac{\partial^2 \Omega}{\partial \beta^2} - \tanh \alpha \frac{\partial \Omega}{\partial \alpha} + \tan \beta \frac{\partial \Omega}{\partial \beta} + k^2 a^2 (\sinh^2 \alpha + \sin^2 \beta)\, \Omega = 0,$$

and there are elementary solutions of the form $\Omega = A(\alpha) B(\beta)$ where

$$\left.\begin{aligned} \frac{d^2 A}{d\alpha^2} - \tanh \alpha \,.\, \frac{dA}{d\alpha} + (\lambda + a^2 k^2 \sinh^2 \alpha) A &= 0 \\ \frac{d^2 B}{d\beta^2} + \tan \beta \,.\, \frac{dB}{d\beta} + (a^2 k^2 \sin^2 \beta - \lambda) \ B &= 0 \end{aligned}\right\} \text{...(194)}.$$

When Ω is independent of t, $k = 0$ and the elementary solutions are of the form

$$\Omega = \int P_n (\xi)\, d\xi \int P_n (\eta)\, d\eta, \qquad \xi = \cosh \alpha, \ \eta = \cos \beta \ \text{...(195)}$$

for prolate spheroids, and of the form

$$\Omega = \int P_n (\xi)\, d\xi \int P_n (\eta)\, d\eta, \qquad \xi = i \sinh \alpha, \ \eta = \sin \beta \text{...(196)}$$

for oblate spheroids. In either of these solutions a function P_n can be replaced by Q_n. The corresponding solutions of Laplace's equation are of the type

$$V = [A P_n (\xi) + B Q_n (\xi)] [C P_n (\eta) + D Q_n (\eta)] \text{...(197)},$$

where A, B, C, D are arbitrary constants.

* *Phil. Trans.* (1880), p. 138.

Examples. 1. Prove that

$$[(\cosh a \cos \beta - \cos \gamma)^2 + \sinh^2 a \sin^2 \beta]^{-\frac{1}{2}}$$
$$= \tfrac{1}{2} \sum_0^\infty (2n+1) Q_n (\cosh a) P_n (\cos \beta) P_n (\cos \gamma).$$

(C. Neumann.)

2. Prove that a function $B(\beta)$ which satisfies the differential equation (192) and is zero for $\beta = 0$, $\beta = \pi$ is a solution of the homogeneous integral equation

$$B(\theta) = \mu \sin^2 \theta \int_0^\pi e^{-ika \cos \theta \cos \beta} B(\beta) \sin \beta \, d\beta \qquad R(ik) \geqslant 0,$$

where μ is determined by the condition that the integral equation should possess a continuous solution which is not identically zero.

(M. Abraham.)

3. If $A(a)$ be defined by the equation

$$A(a) = \sinh^2 a \int_0^\pi e^{-ika \cosh a \cos \beta} B(\beta) \sin \beta \, d\beta,$$

it satisfies the differential equation (192). A second solution of this equation is given by

$$\bar{A}(a) = \sinh^2 a \int_0^\infty e^{-ika \cosh a \cosh \xi} A(\xi) \sinh \xi \, d\xi \qquad R(ik) > 0,$$

and is suitable for the representation of divergent waves.

(M. Abraham.)

§ 34. Paraboloidal coordinates.

If we write

$$z + i\rho = (\alpha_0 + i\beta_0)^2, \quad \alpha_0^2 = -\alpha, \quad \beta_0^2 = \beta$$

so that the transformation is

$$z = -\alpha - \beta, \quad \rho = 2\sqrt{-\alpha\beta} \quad \text{............(198)},$$

the differential equation (143) becomes*

$$\alpha \frac{\partial^2 W}{\partial \alpha^2} + (m+1)\frac{\partial W}{\partial \alpha} - \beta \frac{\partial^2 W}{\partial \beta^2} - (m+1)\frac{\partial W}{\partial \beta} - (\alpha - \beta)\frac{1}{c^2}\frac{\partial^2 W}{\partial t^2} = 0,$$

and is satisfied by†

$$W = A(\alpha) B(\beta) e^{\pm ikct},$$

* Cf. H. J. Sharpe, *Quarterly Journal*, Vol. 15 (1878); *Proc. Camb. Phil. Soc.* Vol. 10 (1899), p. 101; Vol. 13 (1905), p. 133; Vol. 15 (1909), p. 190; H. Lamb, *Proc. London Math. Soc.* Ser. 2, Vol. 4 (1907), p. 190.

† The existence of elementary solutions for the paraboloid and certain other surfaces is established in Bôcher's *Die Reihenentwickelungen der Potentialtheorie*. See the table on pp. 256–7.

if
$$\left.\begin{aligned}\alpha \frac{d^2A}{d\alpha^2} + (m+1)\frac{dA}{d\alpha} - (h - k^2\alpha) A = 0 \\ \beta \frac{d^2B}{d\beta^2} + (m+1)\frac{dB}{d\beta} - (h - k^2\beta) B = 0\end{aligned}\right\}\text{......(199)},$$
where h is arbitrary.

Putting $2ikn = ik(m+1) - h$ we find that the differential equations are satisfied by putting
$$A = e^{-ik\alpha} F_m{}^n (2ik\alpha), \quad B = e^{-ik\beta} F_m{}^n (2ik\beta),$$
where $F_m{}^n(s)$ satisfies the differential equation*
$$s\frac{d^2F}{ds^2} + (m+1-s)\frac{dF}{ds} + nF = 0 \text{.........(200)}.$$

When n is a positive integer, one solution of this equation is furnished by Sonin's polynomial† $T_m{}^n(s)$, which may be defined with the aid of the expansion
$$(1+t)^{-m-1} e^{\frac{st}{1+t}} = \sum_{n=0}^{\infty} \Gamma(m+n+1)\, t^n T_m{}^n(s) \text{...(201)}.$$

A few properties of this function are given here for the sake of reference.
$$T_m{}^n(s) = \frac{s^n}{\Gamma(m+n+1)\,\lfloor n} - \frac{s^{n-1}}{\Gamma(m+n)\,\lfloor n-1\,\lfloor 1} + \frac{s^{n-2}}{\Gamma(m+n-1)\,\lfloor n-2\,\lfloor 2} - \dots \quad \text{...(202)},$$
$$\left.\begin{aligned}\int_0^\infty e^{-s} s^m T_m{}^n(s) T_m{}^\nu(s)\, ds &= 0 & \nu \neq n \\ &= \frac{1}{\Gamma(n+1)\,\Gamma(m+n+1)} & \nu = n\end{aligned}\right\}\text{...(203)},$$
$$\frac{d^p}{ds^p} T_m{}^n(s) = T_{m+p}^{n-p}(s) \text{ ....(204)},$$
$$\frac{d^p}{ds^p}[s^m T_m{}^n(s)] = s^{m-p} T_{m-p}^{n}(s) \text{ ....(205)},$$

* This is a slight modification of Weiler's canonical form for an equation of Laplace's type, *Crelle's Journal*, Bd. 51 (1856), p. 105. The equation is discussed for real values of m and n by O. Schlömilch, *Höheren Analysis*, Bd. 2 (1874), p. 517.

† *Math. Ann.* Bd. 16. Further properties of the function are given by L. Gegenbauer, *Wien. Ber.* (1887), p. 274, who proves that the roots of the equation $T_m{}^n(s) = 0$, considered as an equation of the nth degree in s, are all real, positive and unequal. This is a generalisation of the result obtained by Laguerre for the case $m = 0$. A geometrical proof has been given by Bôcher, *Proc. of the Amer. Acad. of Arts and Sciences*, Vol. 40 (1904).

$$s^m T_m^n(s) = \frac{(-1)^n e^s}{\Gamma(m+n+1)\,\Gamma(n+1)} \frac{d^n}{ds^n}[e^{-s} s^{m+n}] \quad \text{...(206)},$$

$$T_m^n(s) = \frac{1}{\Gamma(m+\frac{1}{2})} \int_0^\pi T_{-\frac{1}{2}}^n (s\cos^2\psi) \sin^{2m}\psi \,.\, d\psi \quad \text{...(207)}.$$

Equations (201), (202) and (203) were given by Abel* and Murphy† for the case $m=0$: the polynomial is then equivalent to the polynomial of Tchebycheff‡ and Laguerre§ which occurs in the theory of interpolation and also in the theory of continued fractions. When $m=\pm\frac{1}{2}$, the polynomial can be expressed in terms of the polynomial $U_n(x)$ discussed by Tchebycheff|| and Hermite¶, or in terms of the function of the parabolic cylinder, discussed by Weber**, Whittaker†† and others‡‡.

The above analysis indicates the existence of a wave-function of the form

$$\Omega = e^{ik(z\pm ct)\pm im\phi}\, T_m^n(2ik\alpha)\, T_m^n(2ik\beta)\, \rho^m \quad \text{...(208)}.$$

This function can be expressed as an integral of the form used in § 5, we have in fact the equation

$$(k\rho)^m T_m^n(2ik\alpha)\, T_m^n(2ik\beta)$$
$$= \frac{(-1)^n}{2\pi\Gamma(m+n+1)} \int_0^{2\pi} e^{k\rho e^{i\gamma}}\, T_0^n[-2k(\rho\cos\gamma + iz)]\, e^{-im\gamma} d\gamma \quad \text{......(209)},$$

from which the required representation can be immediately derived. In this formula m is either zero or a positive integer.

The convergence of a series of terms of type (208) in which

* *Mémoires de mathématique* par N. H. Abel, Paris (1826); *Oeuvres*, Sylow and Lie, t. 2.

† *Cambr. Phil. Trans.* (1833).

‡ *Mém. de l'Acad. de St Pétersbourg* (1860).

§ *Bull. de la Soc. math. de France*, t. 7 (1879); *Oeuvres de Laguerre*, t. 1, p. 428.

|| *Loc. cit.* See also Sturm, *Liouville's Journal*, Vol. 1.

¶ *Comptes Rendus*, t. 58 (1864), p. 93. The Hermite functions have been generalised by Curzon, *Proc. London Math. Soc.* Vol. 13 (1914), p. 417. The generalised functions are intimately connected with the functions considered here.

** *Math. Ann.* Bd. 1 (1869), p. 1.

†† *Proc. London Math. Soc.* Ser. 1, Vol. 35 (1903), p. 417.

‡‡ Baer, *Diss. Cüstrin* (1883); Häntzschel, *Zeitschr. für Math.* Bd. 33 (1888); Adamoff, *Annales de St Pétersbourg*, t. 5 (1906); G. N. Watson, *Proc. London Math. Soc.* Ser. 2, Vol. 8 (1910), p. 393.

n takes different integral values can be partially discussed with the aid of the equation

$$T_m^n(ix)\,T_m^n(-ix)$$
$$=\left[\frac{1}{\Gamma(n+1)\,\Gamma(m+1)}\right]^2\left[1+\frac{n(m+n+1)}{1\,(m+1)^2(m+2)}x^2\right.$$
$$\left.+\frac{n(n-1)(m+n+1)(m+n+2)}{1.2\,(m+1)^2(m+2)^2(m+3)(m+4)}x^4+\dots\right]\quad\dots\dots(210),$$

which shows that the modulus of $T_m^n(ix)$ increases with x. Hence if a series of terms of type (208) converges absolutely for any given value of α, it converges absolutely for all smaller values of α.

For a fuller discussion of the convergence it would be useful to have an asymptotic expression for $T_m^n(s)$ when n is large. Suitable asymptotic expressions have already been found for the case $m=\pm\frac{1}{2}$ by Adamoff and Watson.

The differential equation (200) has been studied for general values of m and n by Pochhammer*, Jacobstahl†, Whittaker‡ and Barnes§. It usually possesses two distinct solutions which can be expanded in power series converging for all finite values of s. If, however, m and n are positive integers, there is only one solution which can be represented by a convergent power series in s, the other may be defined by the equation

$$U_m^n(s)=\frac{1}{\Gamma(n+1)}e^s\int_0^\infty e^{-\sigma}\sigma^n(s-\sigma)^{-m-n-1}d\sigma\dots(211):$$

it contains a logarithmic term. For negative integral values of n we may adopt the definition

$$U_m^n(s)=\frac{d^{-n-1}}{ds^{-n-1}}\{e^s s^{-m-n-1}\}\quad\dots\dots\dots\dots(212).$$

It should be noticed that when $|\alpha|$ is large, $U_m^n(2ik\alpha)$ has an asymptotic expansion of which the first term is

$$(2ik\alpha)^{-m-n-1}e^{2ik\alpha}.$$

The solutions of type

$$e^{ik(z+ct)\pm im\phi}\,U_m^n(2ik\alpha)\,T_m^n(2ik\beta)$$

* *Math. Ann.* Vol. 36, p. 84; Vol. 46, p. 584. † *Ibid.* Vol. 56, p. 129.
‡ *Bull. Amer. Math. Soc.* (1904).
§ *Cambr. Phil. Trans.* Vol. 20 (1906), p. 253.

are consequently suitable when α is large for the representation of waves diverging to infinity in the positive direction of the axis of z.

It may be worth while to mention here that the functions $T_m^n(s)$, $U_m^n(s)$ both satisfy Gegenbauer's difference equations

$$\left.\begin{aligned}
F_m^{n-1}(s) &= (m+n)\,F_{m+1}^{n-1}(s) + F_{m+1}^{n-2}(s)\\
n(m+n)\,F_m^n(s) - \{s-(m+2n-1)\}\,F_m^{n-1}(s) + F_m^{n-2}(s) &= 0\\
(n-1)\,F_m^{n-1}(s) &= \{s-(m+n-1)\}\,F_{m+1}^{n-2}(s) - F_{m+1}^{n-3}(s)\\
(n-1)\,F_m^{n-1}(s) &= \{s-(m+1)\}\,F_{m+1}^{n-2}(s) - s\,F_{m+2}^{n-3}(s)
\end{aligned}\right\} \quad \ldots\ldots(213).$$

The function $U_m^n(s)$ also satisfies an equation analogous to (204).

§ 35. Relations between different solutions.

Many useful formulae may be obtained by expanding known wave-functions in series of elementary wave-functions of type (208) and by identifying our elementary wave-functions with certain definite integrals which are known to represent wave-functions. For instance, we have the expansion $|\tan \tfrac{1}{2}\omega| < 1$

$$e^{ikz\cos\omega} J_m(k\rho\sin\omega)$$

$$= \left(k\rho\tan\frac{\omega}{2}\right)^m e^{ikz}\sec^2\frac{\omega}{2}\sum_{n=0}^{\infty}(-1)^n\, n!\,(m+n)!$$

$$\times \tan^{2n}\frac{\omega}{2}\, T_m^n(2ik\alpha)\, T_m^n(2ik\beta) \quad \ldots\ldots\ldots\ldots\ldots\ldots(214)$$

which enables us to represent a plane wave with the aid of a double series of solutions of the form (208).

Further identities may be obtained by deriving wave-functions from Cunningham's solutions* of the equations

$$\frac{\partial u}{\partial \tau} = \frac{\partial^2 u}{\partial x^2}, \qquad \frac{\partial u}{\partial \tau} = \frac{\partial^2 u}{\partial x^2} + \frac{\partial^2 u}{\partial y^2} \quad \ldots\ldots\ldots(215).$$

The first equation possesses the *polynomial solutions*

$$\tau^{\frac{n}{2}}\, T_{-\frac{1}{2}}^n\left(-\frac{x^2}{4\tau}\right), \qquad x\tau^{\frac{n-1}{2}}\, T_{\frac{1}{2}}^n\left(-\frac{x^2}{4\tau}\right) \quad \ldots\ldots(216)$$

* *Proc. Roy. Soc.* Ser. A, Vol. 81 (1908), p. 310. See also Wera Lebedeff, *Diss. Göttingen* (1906); *Math. Ann.* (1907). The first result is given by Appell, *Liouville's Journal*, Ser. 4, t. 8 (1892), p. 187.

and also the solutions

$$\tau^{-\frac{n+1}{2}} e^{-\frac{x^2}{4\tau}} T^n_{-\frac{1}{2}}\left(\frac{x^2}{4\tau}\right), \qquad x\tau^{-\frac{n+2}{2}} e^{-\frac{x^2}{4\tau}} T^n_{\frac{1}{2}}\left(\frac{x^2}{4\tau}\right) \text{ ...(217).}$$

The second equation possesses the *polynomial solutions* of type

$$\tau^\nu \rho^m T_m{}^\nu \left(-\frac{\rho^2}{4\tau}\right) \sin m(\phi - \phi_0) \text{ ....(218)}$$

and also the solutions of type

$$\tau^{-\nu-m-1} \rho^m e^{-\frac{\rho^2}{4\tau}} T_m{}^\nu \left(\frac{\rho^2}{4\tau}\right) \sin m(\phi - \phi_0) \text{ ...(219).}$$

Wave-functions may be derived from these solutions by the method of § 13.

We add here a few relations which are obtained by expressing the solutions thus formed in terms of old solutions.

$$e^{\sqrt{\xi\eta} e^{i\omega}} T_0{}^n [\xi + \eta - 2\sqrt{\xi\eta} \cos\omega]$$

$$= \sum_{m=-n}^{\infty} (-1)^n \Gamma(m+n+1) \xi^{\frac{m}{2}} \eta^{\frac{n}{2}} T_m{}^n(\xi) T_m{}^n(\eta) e^{im\omega} \text{...(220),}$$

$$\frac{(1-\lambda)^n}{n!} e^{\lambda s} = \sum_{m=-n}^{\infty} s^m \lambda^{m+n} T_m{}^n(s) \text{ ....(221),}$$

$$\tau^{-\nu-m-1} \rho^m e^{-\frac{\rho^2}{4\tau}} T_m{}^n\left(\frac{\rho^2}{4\tau}\right)$$

$$= \frac{(-1)^n 2^{m+1}}{\Gamma(m+n+1)\, n!} \int_0^\infty e^{-\lambda^2\tau} J_m(\lambda\rho) \lambda^{2n+m+1} d\lambda \text{......(222),}$$

$$2^{m+1} \lambda^{2n+m} K_m(\lambda\rho)$$

$$= \Gamma(n+1)\Gamma(n+m+1)\rho^m \int_0^\infty e^{-\lambda^2\tau - \frac{\rho^2}{4\tau}} T_m{}^n\left(\frac{\rho^2}{4\tau}\right) \tau^{-m-n-1} d\tau$$

$$\text{......(223).}$$

The proofs of these are left to the reader.

Prof. G. D. Birkhoff has remarked to me that the differential equation (200) can be regarded as a limiting case of the hypergeometric equation when two of the singularities coincide at infinity*, consequently many properties of the solutions can be derived from known properties of hypergeometric functions†. It should be noticed that when W is independent of t there

* Cf. Bôcher, *Die Reihenentwickelungen der Potentialtheorie*, p. 137.

† This method was used in a particular case by Kummer, *Crelle's Journal*, Bd. 15 (1836), p. 138.

are elementary solutions of equation (143) of the form $W = AB$ where

$$A = \frac{J_m(2i\sqrt{h\alpha})}{(2i\sqrt{h\alpha})^m}, \qquad B = \frac{J_m(2i\sqrt{h\beta})}{(2i\sqrt{h\beta})^m}.$$

We thus obtain elementary solutions of Laplace's equation of the form

$$J_m(\lambda\sqrt{\alpha})\, J_m(\lambda\sqrt{\beta}) \cos m(\phi - \phi_0) \quad \text{.........(224)},$$

where m, λ, ϕ_0 are arbitrary parameters.

§ 36. Toroidal coordinates.

If we put

$$x = \rho\cos\phi, \quad y = \rho\sin\phi, \quad z = \zeta\cosh\omega, \quad ct = \zeta\sinh\omega \quad \text{...(225)},$$

$$\rho + i\zeta = a\coth\frac{\sigma - i\psi}{2},$$

$$\rho = \frac{a\sinh\sigma}{\cosh\sigma - \cos\psi}, \qquad \zeta = \frac{a\sin\psi}{\cosh\sigma - \cos\psi} \quad \text{...(226)},$$

the wave-equation becomes

$$\frac{\partial}{\partial\sigma}\left\{\frac{\sinh\sigma\sin\psi}{(\cosh\sigma - \cos\psi)^2}\frac{\partial u}{\partial\sigma}\right\} + \frac{\partial}{\partial\psi}\left\{\frac{\sinh\sigma\sin\psi}{(\cosh\sigma - \cos\psi)^2}\frac{\partial u}{\partial\psi}\right\}$$

$$+ \frac{\sin\psi}{\sinh\sigma(\cosh\sigma - \cos\psi)^2}\frac{\partial^2 u}{\partial\phi^2} - \frac{\sinh\sigma}{\sin\psi(\cosh\sigma - \cos\psi)^2}\frac{\partial^2 u}{\partial\omega^2} = 0$$

$$\text{......(227)}.$$

This is satisfied by

$$u = F(\sigma)\, G(\psi)(\cosh\sigma - \cos\psi)\, e^{\pm k\omega}\cos m(\phi - \phi_0) \quad \text{...(228)}$$

if

$$\left.\begin{aligned} &\operatorname{cosech}\sigma\frac{d}{d\sigma}\left(\sinh\sigma\frac{dF}{d\sigma}\right) - \left\{n(n+1) + \frac{m^2}{\sinh^2\sigma}\right\}F = 0 \\ &\operatorname{cosec}\psi\frac{d}{d\psi}\left(\sin\psi\frac{dG}{d\psi}\right) + \left\{n(n+1) - \frac{k^2}{\sin^2\psi}\right\}F = 0 \end{aligned}\right\}$$

$$\text{......(229)}.$$

Hence we obtain wave-functions of the form

$$(\cosh\sigma - \cos\psi)\, P_n^m(\cosh\sigma)\, P_n^k(\cos\psi)\, e^{\pm k\omega}\cos m(\phi - \phi_0)$$

$$\text{......(230)}.$$

Other solutions of the wave-equation may be obtained by replacing the functions P_n^m, P_n^k by Q_n^m, Q_n^k.

Many useful formulae may be obtained by expanding particular wave-functions in series of wave-functions of type

(230). The expansion of unity, for instance, gives rise to Neumann's expansion

$$\frac{1}{\cosh\sigma-\cos\psi}=\sum_{n=0}^{\infty}(2n+1)\,Q_n(\cosh\sigma)\,P_n(\cos\psi)\ \ldots(231).$$

It should be noticed that when we make the substitution (225) the wave-equation becomes

$$\frac{\partial^2 u}{\partial\rho^2}+\frac{1}{\rho}\frac{\partial u}{\partial\rho}+\frac{1}{\rho^2}\frac{\partial^2 u}{\partial\phi^2}+\frac{\partial^2 u}{\partial\zeta^2}+\frac{1}{\zeta}\frac{\partial u}{\partial\zeta}-\frac{1}{\zeta^2}\frac{\partial^2 u}{\partial\omega^2}=0\ \ldots(232):$$

it thus possesses elementary solutions of the forms

$$K_p(\lambda\zeta)\,J_m(\lambda\rho)\,e^{\pm p\omega}\cos m(\phi-\phi_0)\ \ \ldots\ldots(233),$$

$$J_p(i\lambda\zeta)\,J_m(\lambda\rho)\,e^{\pm p\omega}\cos m(\phi-\phi_0)\ldots\ldots\ldots(234).$$

The expression of solutions of type (230) in terms of the solutions just found leads to some interesting identities. Thus we have the equation

$$\int_0^\infty K_p(\lambda\zeta)\,J_m(\lambda\rho)\,J_n(\lambda a)\,\lambda^{p+m-n+1}\,d\lambda$$

$$=2^{p+m-n-1}a^{n-p-m-2}\frac{\Gamma(p+m+1)\,\Gamma(p+1)\,\Gamma(m+1)}{\Gamma(n+1)}$$

$$\times(\cosh\sigma-\cos\psi)P^{-p}_{p+m-n}(\cos\psi)\,P^{-m}_{p+m-n}(\cosh\sigma)$$

$$p>-1,\quad m>-1,\quad p+m>-1\ \ldots\ldots\ldots(235).$$

Many important formulae connected with Bessel functions are simply particular cases of this one*. It should be remembered that

$$K_{\frac{1}{2}}(x)=\sqrt{\frac{\pi}{2x}}\,e^{-x}=K_{-\frac{1}{2}}(x)\ \ \ldots\ldots\ldots(236).$$

The corresponding integral in which $K_p(\lambda\zeta)$ is replaced by $J_{-p}(\lambda\xi)$ can also be evaluated in terms of Legendre functions, but the formulae are more complicated. The case $p=-m$ is discussed by Macdonald†.

It should be noticed that if we write

$$\left.\begin{array}{ll}\cosh(\alpha-\omega)=i\cot\psi, & \cos(\beta-\phi)=\coth\sigma\\ \sinh(\alpha-\omega)=i\,\mathrm{cosec}\,\psi, & \sin(\beta-\phi)=\pm i\,\mathrm{cosech}\,\sigma\end{array}\right\}\ldots(237),$$

three relations of type

$$\frac{\partial(\alpha,\beta)}{\partial(x,y)}=\pm\frac{i}{c}\frac{\partial(\alpha,\beta)}{\partial(z,t)}\ \ \ldots\ldots\ldots\ldots\ldots(238)$$

* See, for instance, the formulae given by H. M. Macdonald, *Proc. London Math. Soc.* Ser. 2, Vol. 7, p. 147, and by the author, *ibid.* Vol. 12, *Abstracts*.

† *Loc. cit.* p. 142.

are satisfied and so the functions α, β can be used to obtain an electromagnetic field by the method of § 5. It is easy to verify that the function

$$u = (\cosh\sigma - \cos\psi) f(\alpha, \beta) \quad \text{............(239)}$$

satisfies the wave-equation, f being an arbitrary function.

We add here a few formulae for $P_n^m(\cosh\sigma)$, $Q_n^m(\cosh\sigma)$; these and other formulae will be found in the memoirs of Dr Hobson and Dr Barnes to which we have already referred.

$$P_n^m(\cosh\sigma) = \frac{1}{\Gamma(1-m)} \coth^m \frac{\sigma}{2} \times F\left\{-n, n+1;\ 1-m;\ -\sinh^2\frac{\sigma}{2}\right\}$$
$$= \frac{2^{2m}}{\Gamma(1-m)} (1-e^{-2\sigma})^{-m} e^{-(n+1)\sigma} \times F\{\tfrac{1}{2}-m, 1+n-m;\ 1-2m;\ 1-e^{-2\sigma}\} \quad \sigma > 0,$$

$$Q_n^m(\cosh\sigma) = (-1)^m \sqrt{\pi} \frac{\Gamma(m+n+1)}{\Gamma(n+\frac{3}{2})} (1-e^{-2\sigma})^m e^{-(n+1)\sigma} \times F\{m+\tfrac{1}{2}, n+m+1;\ n+\tfrac{3}{2}, e^{-2\sigma}\} \quad \sigma > 0.$$

Various asymptotic expansions for these functions are given by the authors just named and by Dr Nicholson*.

It should be mentioned that the solutions of the wave-equation that have just been obtained are not directly useful for the treatment of the boundary problems of mathematical physics. They may, however, be used to construct useful solutions of the equation $\Delta u + k^2 u = 0$ by means of various artifices. If, for instance, we multiply one of our wave-functions by e^{ikct} and integrate with regard to t between z and ∞, the resulting function will often be a solution of $\Delta v + k^2 v = 0$. This may be illustrated by taking the wave-function

$$u = J_0(i\lambda\zeta) J_m(\lambda\rho) \cos m(\phi - \phi_0)$$

and using the formula

$$\int_z^\infty J_0\left(\lambda\sqrt{c^2t^2 - z^2}\right) e^{ikct}\, dt = \frac{ie^{iz\sqrt{k^2-\lambda^2}}}{c\sqrt{k^2-\lambda^2}} \quad k^2 > \lambda^2, \quad z > 0$$
$$= \frac{e^{-z\sqrt{\lambda^2-k^2}}}{c\sqrt{\lambda^2-k^2}} \quad \lambda^2 > k^2 \quad \text{......(240).}$$

* *British Association Reports*, Winnipeg (1909), p. 391.

The integration can be taken between other limits in certain cases; for instance, the integral

$$v = \int_{\sqrt{z^2+(\rho-a)^2}}^{\sqrt{z^2+(\rho+a)^2}} \frac{e^{iktc}\,dt}{\sqrt{\{(\rho+a)^2+z^2-c^2t^2\}\{c^2t^2-z^2-(\rho-a)^2\}}} \quad \text{......(241)}$$

represents the solution of $\Delta v + k^2 v = 0$ corresponding to a circular ring of sources. In this case our wave-function u is a constant multiple of $\cosh\sigma - \cos\psi$.

The theory of electrical oscillations on a conducting anchor ring has been treated by H. C. Pocklington*, W. McF. Orr† and Lord Rayleigh‡, without the use of toroidal coordinates; the results are of course only approximate.

§ 37. Solutions of Laplace's equation.

If we put

$$\rho = \frac{a\sinh\sigma}{\cosh\sigma - \cos\psi}, \quad z = \frac{a\sin\psi}{\cosh\sigma-\cos\psi} \quad \text{...(242)},$$

the angle ψ may be interpreted as the angle which two fixed points A, B whose coordinates are $z=0$, $\rho = \pm a$, subtend at a point $P\ (\rho, z)$; the quantity σ may be interpreted as $\log\dfrac{PA}{PB}$. The surfaces $\psi =$ const. are spheres having a real circle ($\rho = a$, $z=0$) in common, the surfaces $\sigma =$ const. are anchor rings.

If we use the toroidal coordinates σ, ψ, ϕ, Laplace's equation becomes§

$$\Delta u \equiv \frac{\partial}{\partial\sigma}\left\{\frac{\sinh\sigma}{\cosh\sigma-\cos\psi}\frac{\partial u}{\partial\sigma}\right\} + \frac{\partial}{\partial\psi}\left\{\frac{\sinh\sigma}{\cosh\sigma-\cos\psi}\frac{\partial u}{\partial\psi}\right\} + \frac{1}{\sinh\sigma(\cosh\sigma-\cos\psi)}\frac{\partial^2 u}{\partial\phi^2} = 0$$

* *Proc. Camb. Phil. Soc.* Vol. 9 (1897), p. 324.

† *Phil. Mag.* Vol. 6 (1903), p. 667.

‡ *Proc. Roy. Soc.* Ser. A, Vol. 87 (1912), p. 93. See also C. W. Oseen, *Phys. Zeitschr.*, Dec. 1st (1913); *Arkiv för Mat. Ast. och Fysik*, Bd. 9 (1913).

§ B. Riemann, *Partielle Differentialgleichungen*, Hattendorf's edition (1861); C. Neumann, *Theorie der Elektricitäts- und Wärme-Vertheilung in einem Ringe*, Halle (1864); W. M. Hicks, *Phil. Trans.* (1881), p. 609; A. B. Basset, *Amer. Journ.* Vol. 15, *Hydrodynamics*, Vol. 2. For an alternative method see F. H. Safford, *Annals of Mathematics*, Vol. 12 (1898), p. 27.

and possesses solutions of the form

$$u = (\cosh\sigma - \cos\psi)^{\frac{1}{2}} \cos n(\psi - \psi_0) \cos m(\phi - \phi_0) \begin{matrix} P^m_{n-\frac{1}{2}}(\cosh\sigma) \\ Q^m_{n-\frac{1}{2}}(\cosh\sigma) \end{matrix} \quad \text{......(243)},$$

which are suitable for the treatment of problems connected with the anchor ring, circular disc and spherical bowl*.

For problems connected with two spheres bipolar coordinates may be used; the appropriate substitution is†

$$\rho = \frac{a\sin\psi}{\cosh\sigma - \cos\psi}, \quad z = \frac{a\sinh\sigma}{\cosh\sigma - \cos\psi} \quad \text{...(244)}.$$

The surfaces $\sigma = \text{const.}$ are now coaxal spheres with imaginary common circle. The radius of the sphere $\sigma = \sigma_0$ is $a\,|\operatorname{cosech}\sigma_0|$ and the distance of its centre from the origin is $a\,|\coth\sigma_0|$.

The ratio of the distances of a point from the limiting points of the system of coaxal spheres is e^σ and the angle between the radii from these points is ψ.

The appropriate solutions of Laplace's equation are now of the type‡

$$u = (\cosh\sigma - \cos\psi)^{\frac{1}{2}} [A\cosh(n+\tfrac{1}{2})\sigma + B\sinh(n+\tfrac{1}{2})\sigma]$$
$$\times \cos m(\phi - \phi_0)\,[f P_n^m(\cos\psi) + g Q_n^m(\cos\psi)] \text{...(245)}.$$

It should be noticed that when we are using toroidal coordinates the function

$$u = (\cosh\sigma - \cos\psi)^{\frac{1}{2}} f\left[\phi \pm i\log\tanh\frac{\sigma}{2}\right] \cos\tfrac{1}{2}(\psi - \psi_0) \quad \text{......(246)}$$

satisfies Laplace's equation and that when we use bipolar coordinates the corresponding solution is

$$(\cosh\sigma - \cos\psi)^{\frac{1}{2}} f\left[\phi \pm i\log\tan\frac{\psi}{2}\right] \cosh\tfrac{1}{2}(\sigma - \sigma_0) \text{...(247)}.$$

* See for instance E. W. Hobson, *Cambr. Phil. Trans.* Vol. 18 (1899); C. W. Oseen, *Arkiv för Matematik*, Bd. 2, No. 5; H. C. Pocklington, *Phil. Trans.* A, Vol. 186 (1895), p. 603.

† W. Thomson (Lord Kelvin), *Liouville's Journal* (1847).

‡ G. B. Jeffery, *Proc. Roy. Soc.* Ser. A, Vol. 87 (1912); G. R. Dean, *Phys. Review* (1912); G. Darboux, *Bull. des Sciences math.* t. 31 (1907), p. 17. Another method of dealing with problems connected with two spheres is described by A. Guillet and M. Aubert, *Journal de Physique*, t. 3 (1913).

It should be mentioned here that other simple solutions may be obtained by using the formulae

$$Q^{\frac{1}{2}}_{n-\frac{1}{2}}(\cosh\sigma) = i\sqrt{\frac{\pi}{2\sinh\sigma}}\,e^{-n\sigma},$$

$$P^{\frac{1}{2}}_{n-\frac{1}{2}}(\cos\psi) = \sqrt{\frac{2}{\pi\sin\psi}}\,\cos n\psi,$$

$$P^{\frac{1}{2}}_{n-\frac{1}{2}}(\cosh\sigma) = -\sqrt{\frac{2}{\pi\sinh\sigma}}\cosh n\sigma.$$

EXAMPLES.

1. If with the notation of § 36 we write

$$\zeta + i\rho = a\cosh(\alpha + i\beta)$$

the wave-equation becomes

$$\frac{\partial^2 u}{\partial \alpha^2} + \frac{\partial^2 u}{\partial \beta^2} + 2\coth 2\alpha\frac{\partial u}{\partial \alpha} + 2\cot 2\beta\frac{\partial u}{\partial \beta} + (\operatorname{cosech}^2\alpha + \operatorname{cosec}^2\beta)\frac{\partial^2 u}{\partial \phi^2} + (\operatorname{sech}^2\alpha - \sec^2\beta)\frac{\partial^2 u}{\partial \omega^2} = 0.$$

Hence show that there are wave-functions of the form

$$u = A(\alpha)\,B(\beta)\,e^{\pm k\omega}\cos m(\phi - \phi_0),$$

where a, k, m and ϕ_0 are arbitrary constants.

2. Prove that if $\rho + iz = f(\alpha + i\beta)$ the wave-equation becomes

$$\frac{\partial^2 u}{\partial \alpha^2} + \frac{\partial^2 u}{\partial \beta^2} + \frac{1}{\rho}\left(\frac{\partial \rho}{\partial \alpha}\frac{\partial u}{\partial \alpha} + \frac{\partial \rho}{\partial \beta}\frac{\partial u}{\partial \beta}\right) + \frac{1}{J^2}\left(\frac{1}{\rho^2}\frac{\partial^2 u}{\partial \phi^2} - \frac{1}{c^2}\frac{\partial^2 u}{\partial t^2}\right) = 0,$$

and obtain elementary solutions of type $A(\alpha)\,B(\beta)\,e^{im\phi \pm ikct}$ when $z + i\rho = a\cosh(\alpha + i\beta)$. Notice that the solutions of equation (188) are not wave-functions, they are analogous to the stream-line functions of hydrodynamics.

CHAPTER VII

HOMOGENEOUS SOLUTIONS OF THE WAVE-EQUATION

§ 38. **The method of Stieltjes***.

Wave-functions which are homogeneous functions of x, y, z, t may be studied with the aid of the substitution

$$\left.\begin{array}{ll} x = s\cos\theta\cos\phi, & z = s\sin\theta\cos\chi \\ y = s\cos\theta\sin\phi, & ict = s\sin\theta\sin\chi \end{array}\right\} \quad \text{......(248).}$$

The wave-equation in these coordinates has the form

$$\frac{\partial^2 u}{\partial s^2} + \frac{3}{s}\frac{\partial u}{\partial s} + \frac{1}{s^2}\frac{\partial^2 u}{\partial \theta^2} + \frac{1}{s^2\cos^2\theta}\frac{\partial^2 u}{\partial \phi^2} + \frac{1}{s^2\sin^2\theta}\frac{\partial^2 u}{\partial \chi^2} + \frac{\cot\theta - \tan\theta}{s^2}\frac{\partial u}{\partial \theta} = 0.$$

Putting $\cos 2\theta = \mu$, we find that there are elementary solutions of degree $2n$ of the form

$$u = s^{2n}\,\Theta(\mu)\,e^{im\phi + ip\chi} \quad \text{..............(249),}$$

if

$$\frac{d}{d\mu}\left\{(1-\mu^2)\frac{d\Theta}{d\mu}\right\} + \Theta\left\{n(n+1) - \frac{m^2}{2(1+\mu)} - \frac{p^2}{2(1-\mu)}\right\} = 0 \quad \text{.........(250).}$$

This equation is satisfied by

$$(1+\mu)^{\frac{m}{2}}(1-\mu)^{\frac{p}{2}} \left(n+1+\frac{m+p}{2}, \frac{m+p}{2} - n, p+1, \frac{1-\mu}{2}\right) \quad \text{.........(251),}$$

with the usual notation of the hypergeometric function †.

* *Comptes Rendus*, t. 95 (1882), p. 901; *Liouville's Journal*, Ser. 4, t. 5 (1889), p. 55. See also Tisserand, *Traité de mécanique céleste*, Paris (1889); H. Bateman, *Proc. London Math. Soc.* Ser. 2, Vol. 3 (1905), p. 111.

† This gives a polynomial if either $n+1+\frac{m+p}{2}$ or $\frac{m+p}{2} - n$ is zero or a negative integer.

It should be noticed that if we write

$$\cos\theta = \sec\alpha,\quad \sin\theta = i\tan\alpha,\quad \sin\chi = \cosh w,\quad \cos\chi = -i\sinh w,$$

$$\mu = \sec^2\alpha + \tan^2\alpha,$$

we obtain *real* wave-functions of the form

$$u = s^{2n} e^{-pw}\,\Theta(\mu)\cos m(\phi - \phi_0) \quad \text{.........(252)},$$

the variable in the hypergeometric function is now $-\tan^2\alpha$. When $m = p$ the equation for Θ is the equation satisfied by the associated Legendre functions. We thus obtain wave-functions of the form

$$u = s^{2n} P_n^m(\cos 2\theta)\, e^{im(\phi+\chi)} \quad \text{.........(253)}.$$

Comparing this with the elementary solution of Laplace's equation in polar coordinates (r, θ, ϕ), we see that if $f(r, \theta, \phi)$ is a solution of Laplace's equation $f(s^2, 2\theta, \phi + \chi)$ is a wave-function. We may thus derive wave-functions from harmonic functions; in particular, the fundamental harmonic function $\frac{1}{r}$ gives us the fundamental wave-function $\frac{1}{s^2} = \frac{1}{x^2 + y^2 + z^2 - c^2 t^2}$. We have already remarked in § 13 that Lord Kelvin's method of inversion may be extended to wave-functions, it is easy to see that the result is an immediate consequence of the fact that the differential equation (250) is unaltered when $-(n+1)$ is written in place of n.

It is easy to see that there are $(n+1)^2$ linearly independent polynomial solutions of degree n, for a general polynomial of degree n contains $\frac{1}{6}(n+1)(n+2)(n+3)$ coefficients and when this is operated on with Ω the vanishing of the resulting polynomial of degree $n-2$ gives $\frac{1}{6}(n-1)n(n+1)$ conditions. The difference between these two numbers is $(n+1)^2$.

A polynomial solution of degree n is given by the integral

$$u = \int_0^{2\pi} (x\cos\alpha + y\sin\alpha + iz)^p (x\sin\alpha - y\cos\alpha - ct)^{n-p} e^{im\alpha}\, d\alpha.$$

A set of $(n+1)^2$ linearly independent polynomials is obtained by allowing m and p to take the values $(0, 1 \ldots n)$. A better set of solutions is obtained by using the integrals of type

$$u = \int_0^\pi [(x - iy) e^{i\alpha} + i(z - ct)]^p \times [(x + iy) e^{-i\alpha} + i(z + ct)]^{n-p} e^{im\alpha} d\alpha.$$

Polynomial solutions may also be obtained by differentiating the fundamental wave-function $\frac{1}{s^2}$ and using generalised inversion*. The polynomial solutions were first discussed by Cayley†. Waelsch‡ has recently studied them from a new point of view.

Example. Prove that when n is a positive integer

$$\frac{1.3\ldots 2n-1}{2.4\ldots 2n} P_n(\cos 2\theta) = \frac{1}{\pi^2}\int_0^\pi \int_0^\pi (\cos\theta\cos\phi + i\sin\theta\cos\chi)^{2n} d\phi\, d\chi.$$

§ 39. The method of Green§.

Homogeneous solutions may also be investigated with the aid of Green's substitution

$$\left.\begin{aligned} x &= s\sin\alpha\sin\beta\cos\phi, \quad & y &= s\sin\alpha\sin\beta\sin\phi \\ z &= s\sin\alpha\cos\beta, & ict &= s\cos\alpha \end{aligned}\right\}\ldots(254).$$

The wave-equation now becomes

$$\frac{\partial^2 u}{\partial s^2} + \frac{3}{s}\frac{\partial u}{\partial s} + \frac{2}{s^2}\cot\alpha\frac{\partial u}{\partial \alpha} + \frac{1}{s^2}\frac{\partial^2 u}{\partial \alpha^2}$$
$$+ \frac{1}{s^2\sin^2\alpha\sin\beta}\frac{\partial}{\partial\beta}\left(\sin\beta\frac{\partial u}{\partial\beta}\right) + \frac{1}{s^2\sin^2\alpha\sin^2\beta}\frac{\partial^2 u}{\partial\phi^2} = 0\ldots(255),$$

and possesses elementary solutions of the form

$$u = s^n A(\alpha) B(\beta)\cos m(\phi - \phi_0),$$

* Cf. F. Didon, *Annales de l'École Normale* (1), t. 5 (1868), p. 229; t. 6 (1869), p. 7; t. 7 (1870), pp. 89, 247; P. Appell, *Rend. Palermo*, t. 36 (1913); K. de Fériet, *Comptes Rendus*, Nov. 17th (1913).

† *Liouville's Journal*, t. 13 (1848); *Phil. Trans.* Vol. (165) II. (1875), p. 675. See also Hermite, *Oeuvres*, t. 2.

‡ *Deutsche Math. Verein*, Bd. 19, p. 90.

§ *Cambr. Phil. Trans.* Vol. 5 (1835), p. 395; *Collected Papers*, p. 187; Cayley, *loc. cit.* See also Heine, *Handbuch der Kugelfunktionen*, Bd. 1, p. 449; *Crelle*, Bd. 60, 61, 62 (1862—1863); E. W. Hobson, *Proc. London Math. Soc.* Ser. 1, Vol. 24, p. 67, Vol. 25; F. G. Mehler, *Progr. Danzig* (1864); *Crelle*, Bd. 66 (1866), p. 161; C. Neumann, *Zeitschr. Math. Phys.* Bd. 12 (1867), p. 116; V. Giulotto, *Gior. d. Mat.* 39 (1901), p. 162.

where

$$\frac{1}{\sin\beta}\frac{d}{d\beta}\left(\sin\beta\frac{dB}{d\beta}\right)+\left[\nu(\nu+1)-\frac{m^2}{\sin^2\beta}\right]B=0,$$

$$\frac{d^2A}{d\alpha^2}+2\cot\alpha\frac{dA}{d\alpha}+\left[n(n+2)-\frac{\nu(\nu+1)}{\sin^2\alpha}\right]A=0.$$

We may thus take

$$\left.\begin{aligned}B&=c_1P_m^\nu(\cos\beta)-c_2Q_m^\nu(\cos\beta)\\ A&=\sqrt{\operatorname{cosec}\alpha}\left[b_1P_{n+\frac{1}{2}}^{\nu+\frac{1}{2}}(\cos\alpha)+b_2Q_{n+\frac{1}{2}}^{\nu+\frac{1}{2}}(\cos\alpha)\right]\end{aligned}\right\}\text{...(256)},$$

where c_1, c_2, b_1, b_2 are arbitrary constants.

§ 40. Wave-functions of degree zero.

If Ω is a wave-function of degree -2, the formulae

$$\left.\begin{aligned}H_x&=y\frac{\partial\Omega}{\partial z}-z\frac{\partial\Omega}{\partial y}, & E_x&=\frac{x}{c}\frac{\partial\Omega}{\partial t}+ct\frac{\partial\Omega}{\partial x}\\ H_y&=z\frac{\partial\Omega}{\partial x}-x\frac{\partial\Omega}{\partial z}, & E_y&=\frac{y}{c}\frac{\partial\Omega}{\partial t}+ct\frac{\partial\Omega}{\partial y}\\ H_z&=x\frac{\partial\Omega}{\partial y}-y\frac{\partial\Omega}{\partial x}, & E_z&=\frac{z}{c}\frac{\partial\Omega}{\partial t}+ct\frac{\partial\Omega}{\partial z}\end{aligned}\right\}\text{...(257)}$$

give a solution of Maxwell's equations.

A homogeneous wave-function of degree -2 can, of course, be derived from a homogeneous wave-function of degree zero. If F is an arbitrary function of two variables subject to suitable restrictions, the integral

$$\Omega=\int_0^{2\pi}F\left[\frac{x\sin\alpha-y\cos\alpha-ct}{x\cos\alpha+y\sin\alpha+iz},\alpha\right]d\alpha\text{......(258)}$$

represents a wave-function of degree zero, and when this is multiplied by $\frac{1}{s^2}$ a wave-function of degree -2 is obtained. We add here a few particular wave-functions of degree zero:

$$f\left(\frac{x\pm iy}{z\pm ct}\right),\quad \tan^{-1}\frac{y}{x},\quad \tanh^{-1}\frac{ct}{z},\quad \log\frac{z^2-c^2t^2}{x^2+y^2}\text{......(259)}.$$

Electromagnetic fields which are derived from this type of wave-function of degree -2 may be generalised by writing $x-\xi(\tau)$, $y-\eta(\tau)$, $z-\zeta(\tau)$, $t-\tau$ instead of x, y, z, t respectively and integrating round a closed contour in the complex τ plane. The integrals thus obtained can generally be evaluated by means of Cauchy's theorem. Many of the results given in the next chapter are suggested at once by this method and may be thoroughly established by a method of direct verification.

It is worthy of note that if we write r in place of ct in a wave-function of degree zero the resulting function is a solution of Laplace's equation $\Delta u = 0$. A general solution of Laplace's equation of degree zero can be derived at once in this way from the first of the solutions (259). We thus obtain Donkin's formula*

$$u = f\left(\frac{x+iy}{z+r}\right) + g\left(\frac{x-iy}{z+r}\right) \quad \text{.........(260)}.$$

A similar result is that if

$$\Omega = F(x, y, z, w, t)$$

is a homogeneous function of degree $-\frac{1}{2}$ satisfying the equation

$$\frac{\partial^2 \Omega}{\partial x^2} + \frac{\partial^2 \Omega}{\partial y^2} + \frac{\partial^2 \Omega}{\partial z^2} + \frac{\partial^2 \Omega}{\partial w^2} - \frac{1}{c^2}\frac{\partial^2 \Omega}{\partial t^2} = 0,$$

and s be written in place of w, the resulting function is a wave-function. Now if $f(x, y, z)$ is a solution of Laplace's equation, the function

$$\Omega = \frac{1}{\sqrt{w-ct}} f\left(\frac{x}{w-ct}, \frac{y}{w-ct}, \frac{z}{w-ct}\right)$$

satisfies the requirements, consequently we may conclude that the function

$$\Omega = \frac{1}{\sqrt{s-ct}} f\left(\frac{x}{s-ct}, \frac{y}{s-ct}, \frac{z}{s-ct}\right) \quad \text{.........(261)}$$

is a wave-function†. Other wave-functions may be derived from this by generalised inversion or by interchanging the variables x, y, z, ict.

* *Phil. Trans.* (1857). This solution may be obtained at once from Jacobi's theorem that if p, q, r are three functions of u which satisfy the equation $p^2 + q^2 + r^2 = 0$ and u is defined by the equation $au = xp(u) + yq(u) + zr(u)$, then an arbitrary function of u is a solution of Laplace's equation, *Werke*, Bd. 2, p. 208. See also Forsyth, *Mess. of Math.* (1898).

† This result is obtained in another way by Pockels, *Über die partielle Differentialgleichung* $\Delta u + k^2 u = 0$. Teubner, Leipzig (1891).

CHAPTER VIII

ELECTROMAGNETIC FIELDS WITH MOVING SINGULARITIES

§ 41. **An electromagnetic field with a simple singularity or electron, first model of a corpuscle*.**

We shall now derive a family of wave-functions from the fundamental wave-function $1/s^2$, where

$$s^2 = [x - \xi(\tau)]^2 + [y - \eta(\tau)]^2 + [z - \zeta(\tau)]^2 - c^2[t - \tau]^2 \quad \ldots(262)$$

and τ is a variable parameter, which is at first independent of x, y, z, t. Using a method invented by Prof. A. W. Conway † we consider the integral

$$\Omega = -\frac{1}{2\pi i}\int \frac{f(\tau)\,d\tau}{s^2},$$

taken round a closed contour in the plane of the complex variable τ. If this contour contains only one root τ of the equation $s^2 = 0$, the value of the integral is

$$\Omega = \frac{f(\tau)}{2\nu},$$

where

$$\nu = \xi'(\tau)(x - \xi) + \eta'(\tau)(y - \eta) + \zeta'(\tau)(z - \zeta) - c^2(t - \tau) \quad \ldots\ldots(263)$$

and τ is the root in question.

* I have ventured to use Johnstone Stoney's term "electron" to denote the simple point singularity and Sir Joseph Thomson's term "corpuscle" to denote the elementary charged particle which has been discovered by experimental work.

† *Proc. London Math. Soc.* Ser. 2, Vol. 1 (1903). Integrals over complex paths had been employed previously in electromagnetic theory by Sommerfeld and other writers.

This function ν vanishes when $x=\xi, y=\eta, z=\zeta, t=\tau$ and so the wave-function Ω has a singularity which moves along the curve Γ represented by

$$x=\xi(\tau), \quad y=\eta(\tau), \quad z=\zeta(\tau) \quad \text{.........(264).}$$

If, moreover, the velocity of this singularity E is always less than the velocity of light, it is easy to see that ν does not vanish for any *real* values of (x, y, z, t) other than those just mentioned.

When the velocity of the singularity E is always less than c, there is only one value of τ less than t for which the equation $s^2=0$ is satisfied: x, y, z, t being supposed to be given.

To prove this we surround each point E on Γ by a sphere of radius $c(t-\tau)$ having E as centre; then it is clear that each sphere lies entirely within the neighbouring one corresponding to a smaller value of τ, provided $\tau<t$ and $d\xi^2+d\eta^2+d\zeta^2<c^2d\tau^2$.

This shows that one and only one of these spheres passes through a given point of space and so *there is only one value of* $\tau<t$ *for which the equation*

$$[x-\xi(\tau)]^2+[y-\eta(\tau)]^2+[z-\zeta(\tau)]^2=c^2(t-\tau)^2 \text{...(265)}$$

is satisfied*.

Now let a point $Q(x, y, z, t)$ move with a velocity less than c along a curve G and let us consider the variation of τ with t. As t increases from t to $t+dt$ the radius of the sphere associated with each point E will increase by cdt and since Q moves a distance less than cdt in the interval dt, its new position will lie within the new sphere associated with the time τ. Consequently the new position of Q lies on a sphere associated with a greater time τ.

Hence if Q moves in any manner with a velocity less than the velocity of light, τ *increases with t.*

Things are quite different when the velocity of E is greater than c. The spheres then have a real envelope and there may be more than one sphere through a given point in space, also τ may sometimes decrease when t increases.

In this case, however, ν vanishes for values of x, y, z, t other than $x=\xi$, $y=\eta$, $z=\zeta$, $t=\tau$, and so Ω has ∞^1 singular lines

* Cf. A. W. Conway, *loc. cit.*; H. Bateman, *Manchester Memoirs* (1910); G. A. Schott, *Electromagnetic Radiation* (1912). The theorem is due to Liénard.

through each point E. These singular lines form a right circular cone whose axis is along E's direction of motion: each singular line is described by a singular point that travels with the velocity of light.

When the velocity of the singularity is less than c we can obtain a solution of Maxwell's equations having the moving singularity by using the potentials*

$$A_x = \frac{e\xi'(\tau)}{4\pi\nu}, \quad A_y = \frac{e\eta'(\tau)}{4\pi\nu}, \quad A_z = \frac{e\zeta'(\tau)}{4\pi\nu}, \quad \Phi = \frac{ec}{4\pi\nu} \qquad \text{......(266)}.$$

It is easy to verify that they satisfy the relation

$$\text{div}\, A + \frac{1}{c}\frac{\partial\Phi}{\partial t} = 0 \qquad \text{..................(267)}.$$

When the electric and magnetic forces are calculated from these potentials with the aid of the formulae

$$H = \text{rot}\, A, \quad E = -\frac{1}{c}\frac{\partial A}{\partial t} - \text{grad}\, \Phi \text{.........(268)}$$

it is found that†

$$H_x = \frac{e}{4\pi}\frac{\partial(\tau, \sigma)}{\partial(y, z)}, \quad E_x = \frac{e}{4\pi c}\frac{\partial(\tau, \sigma)}{\partial(x, t)} \qquad \text{......(269)},$$

....

where

$$\sigma\nu = \xi''(x - \xi) + \eta''(y - \eta) + \zeta''(z - \zeta) - (\xi'^2 + \eta'^2 + \zeta'^2) + c^2 \qquad \text{......(270)}.$$

It is clear from these equations that the magnetic force is

* A. Liénard, *L'éclairage électrique*, Vol. 16 (1898), pp. 5, 53, 106. See also E. Wiechert, *Arch. néerlandaises* (2), Vol. 5 (1900), p. 54; K. Schwarzschild, *Gött. Nachr.* (1903). The potentials are usually written in the form $\Phi = \dfrac{e}{4\pi\left[r\left(1 - \frac{v_r}{c}\right)\right]}$, $A = \dfrac{e[v]}{4\pi c\left[r\left(1 - \frac{v_r}{c}\right)\right]}$, where the square bracket indicates that the quantity enclosed is to be calculated at time $\tau = t - \frac{r}{c}$. Cf. H. A. Lorentz, *The Theory of Electrons*, p. 50. To obtain a model of a corpuscle we must write de instead of e and integrate over a small region.

† These expressions for the components of E and H were communicated to me by Mr R. Hargreaves in 1909; they should be of some historical interest in connection with the general theory of § 5. This was, however, the outcome of some independent work. Cf. *Proc. London Math. Soc.* (2), Vol. 10 (1911), p. 96.

perpendicular to the electric force and also to the radius from the effective position of E; for we have the relations

$$\left.\begin{aligned} \nu\frac{\partial\tau}{\partial x} &= x-\xi, & \nu\frac{\partial\tau}{\partial y} &= y-\eta \\ \nu\frac{\partial\tau}{\partial z} &= z-\zeta, & \nu\frac{\partial\tau}{\partial t} &= -c^2(t-\tau) \end{aligned}\right\} \quad \text{......(271).}$$

It also follows from these relations that τ satisfies the characteristic equation

$$\left(\frac{\partial\tau}{\partial x}\right)^2+\left(\frac{\partial\tau}{\partial y}\right)^2+\left(\frac{\partial\tau}{\partial z}\right)^2=\frac{1}{c^2}\left(\frac{\partial\tau}{\partial t}\right)^2 \quad \text{.........(272).}$$

This is to be expected because, as Jacobi has remarked* for the case of Laplace's equation, the argument τ of an arbitrary function occurring in the solution of a partial differential equation must satisfy the partial differential equation of the characteristics†.

To prove that there is a constant charge e associated with the singularity of our electromagnetic field we shall calculate the integral of the radial component of E over a sphere having the singularity as centre. We have to evaluate the integral

$$\frac{ce}{4\pi}\iint\left(\frac{\partial\sigma}{\partial x}\frac{\partial\tau}{\partial x}+\frac{\partial\sigma}{\partial y}\frac{\partial\tau}{\partial y}+\frac{\partial\sigma}{\partial z}\frac{\partial\tau}{\partial z}-\frac{1}{c^2}\frac{\partial\sigma}{\partial t}\frac{\partial\tau}{\partial t}\right)dS,$$

which is easily transformed into

$$\frac{ce}{4\pi}\iint\frac{c^2-\xi'^2-\eta'^2-\zeta'^2}{\nu^2}\,dS.$$

Transforming the axes so that the axis of z is in the direction of motion of the singularity, we may put

$$\nu=r(v\cos\theta-c),\quad dS=r^2\sin\theta\,d\theta\,d\phi,\quad \xi'^2+\eta'^2+\zeta'^2=v^2,$$

and our integral becomes

$$\frac{ce}{4\pi}\int_0^\pi\int_0^{2\pi}\frac{(c^2-v^2)\sin\theta\,d\theta\,d\phi}{(v\cos\theta-c)^2}=e.$$

* *Werke*, Bd. 2, p. 208; *Crelle's Journal*, Bd. 36 (1848).

† For the general theory of characteristics see Hadamard, *Propagation des Ondes* (1903), Chapters VII. and VIII.; J. Coulon, *Comptes Rendus*, t. 128 (1899), p. 1386; A. V. Bäcklund, *Math. Ann.* Bd. 13 (1878), p. 411; J. Beudon, *Comptes Rendus*, t. 124 (1897), p. 124.

Prof. E. T. Whittaker* has calculated potentials Γ, Π from which A and Φ can be derived by using (7). He finds that

$$\Pi = (0, 0, S), \qquad \Gamma = (0, 0, N),$$

where

$$\left.\begin{aligned} 4\pi S = e \sinh^{-1} \frac{z-\zeta}{\sqrt{(x-\xi)^2+(y-\eta)^2}}, \qquad 4\pi N = -e \tan^{-1}\frac{y-\eta}{x-\xi} \\ 4\pi K = -e \log \sqrt{(x-\xi)^2+(y-\eta)^2} \end{aligned}\right\} \quad \ldots\ldots(273).$$

It may be verified without difficulty that the functions are wave-functions. This result is a particular case of the following general theorem.

If $f(x, y, z)$ is a homogeneous function of degree zero satisfying Laplace's equation $\Delta u = 0$, the function

$$\Omega = f[x-\xi(\tau), y-\eta(\tau), z-\zeta(\tau)] \quad \ldots\ldots(274)$$

is a wave-function.

§ 42. The electromagnetic field due to a moving doublet.

Let us now derive an electromagnetic field by superposing two electromagnetic fields of the type just described wherein the singularities move along the two neighbouring curves

$$x = \xi(\tau), \qquad y = \eta(\tau), \qquad z = \zeta(\tau),$$
$$x = \xi(\tau_1) + \epsilon\alpha(\tau_1), \qquad y = \eta(\tau_1) + \epsilon\beta(\tau_1), \qquad z = \zeta(\tau_1) + \epsilon\gamma(\tau_1),$$

ϵ being a quantity whose square may be neglected.

If τ_1 is defined in terms of x, y, z, t by the equation

$$[x-\xi(\tau_1)-\epsilon\alpha(\tau_1)]^2 + [y-\eta(\tau_1)-\epsilon\beta(\tau_1)]^2 + [z-\zeta(\tau_1)-\epsilon\gamma(\tau_1)]^2 = c^2(t-\tau_1)^2$$

and $\tau_1 = \tau + \epsilon\theta$, we easily find that

$$\nu\theta + \alpha(x-\xi) + \beta(y-\eta) + \gamma(z-\zeta) = 0.$$

Also if ν_1 is the quantity corresponding to ν, we have

$$\nu_1 = \nu + \epsilon[\theta\nu\sigma + \alpha'(x-\xi) + \beta'(y-\eta) + \gamma'(z-\zeta) - \alpha\xi' - \beta\eta' - \gamma\zeta'] = \nu + \epsilon[\theta\nu\sigma + \rho], \text{ say.}$$

Now if $\qquad A_x' = \dfrac{e\{\xi'(\tau_1) + \epsilon\alpha'(\tau_1)\}}{4\pi\nu_1}, \qquad \Phi' = \dfrac{ec}{4\pi\nu_1},$

* *Proc. London Math. Soc.* Ser. 2, Vol. 1.

we find that

$$a_x = \frac{1}{\epsilon}[A_x' - A_x] = \frac{e}{4\pi\nu^2}[\nu\alpha'(\tau) - \rho\xi'(\tau) + \nu\theta\xi''(\tau) - \nu\theta\sigma\xi'(\tau)],$$

$$\phi = \frac{1}{\epsilon}[\Phi' - \Phi] = -\frac{ec}{4\pi\nu^2}[\rho + \nu\theta\sigma].$$

But

$$\nu\alpha' - \rho\xi' + \nu\theta\xi'' - \nu\theta\sigma\xi' \equiv n'(y-\eta) - m'(z-\zeta) - c^2\alpha'(t-\tau)$$
$$+ c^2\alpha - \eta' n - \zeta' m - \sigma\{\alpha(t-\tau) - n(y-\eta) + m(z-\zeta)\},$$

where $\quad l = \beta\zeta' - \gamma\eta', \quad m = \gamma\xi' - \alpha\zeta', \quad n = \alpha\eta' - \beta\xi'.$

Hence we may write

$$\left.\begin{aligned} a_x &= \frac{e}{4\pi}\left[\frac{\partial}{\partial y}\left(\frac{n}{\nu}\right) - \frac{\partial}{\partial z}\left(\frac{m}{\nu}\right) + \frac{\partial}{\partial t}\left(\frac{\alpha}{\nu}\right)\right] \\ \phi &= -\frac{ec}{4\pi}\left[\frac{\partial}{\partial x}\left(\frac{\alpha}{\nu}\right) + \frac{\partial}{\partial y}\left(\frac{\beta}{\nu}\right) + \frac{\partial}{\partial z}\left(\frac{\gamma}{\nu}\right)\right] \end{aligned}\right\} \quad \ldots(275).$$

The electromagnetic field derived from these potentials is due to a moving electric doublet. It should be noticed that we have the relations

$$\left.\begin{aligned} l\alpha + m\beta + n\gamma &= 0 \\ l\xi' + m\eta' + n\zeta' &= 0 \end{aligned}\right\} \quad \ldots\ldots\ldots\ldots\ldots(276).$$

We can write down by analogy the potentials for an electromagnetic field due to a moving magnetic doublet. They are

$$\left.\begin{aligned} a_x &= \frac{\mu}{4\pi}\left[\frac{\partial}{\partial y}\left(\frac{\gamma_0}{\nu}\right) - \frac{\partial}{\partial z}\left(\frac{\beta_0}{\nu}\right) - \frac{\partial}{\partial t}\left(\frac{l_0}{\nu}\right)\right] \\ \phi &= -\frac{\mu c}{4\pi}\left[\frac{\partial}{\partial x}\left(\frac{l_0}{\nu}\right) + \frac{\partial}{\partial y}\left(\frac{m_0}{\nu}\right) + \frac{\partial}{\partial z}\left(\frac{n_0}{\nu}\right)\right] \end{aligned}\right\} \quad \ldots(277),$$

where

$$\left.\begin{aligned} l_0\alpha_0 + m_0\beta_0 + n_0\gamma_0 &= 0 \\ l_0\xi' + m_0\eta' + n_0\zeta' &= 0 \end{aligned}\right\} \quad \ldots\ldots\ldots\ldots(278),$$

and $\alpha_0, \beta_0, \gamma_0, l_0, m_0, n_0$ are functions of τ. When $\alpha, \beta, \gamma, l, m, n$ are functions of τ which are not connected by the relations (276) the potentials (275) can be used to construct an electromagnetic field which must be regarded as that due to an electric doublet and magnetic doublet which move together.

§ 43. **Electromagnetic fields in which singularities are projected from a moving point or curve and travel with the velocity of light.**

We shall now develop some mathematical analysis of considerable interest whose physical significance is not yet fully understood. At first sight it seems appropriate for a discussion of an emission theory of light in which waves in the aether are either produced or guided by small particles which move in straight lines with the velocity c. After further study I have thought it may be useful for a discussion of the question "Has the aether a structure?"

This question has already been raised by Sir Joseph Larmor* and Sir Joseph Thomson†. The latter has, indeed, developed a theory in which the aether has a kind of atomic structure of which the elements are Faraday tubes‡. In the most recent form of the theory it is assumed that the electric and magnetic forces are zero outside the tubes and that a certain amount of work is performed when one corpuscle crosses a tube of force attached to another. In an application of the present analysis to Sir Joseph Thomson's theory the aim would be to build up his discontinuous electromagnetic fields from electromagnetic fields with certain types of singularities, making use of discontinuous definite integrals. To illustrate the possibility of doing this it will be sufficient to mention the definite integral

$$V = \int_0^{2\pi} \left[\frac{d\alpha}{z + ix\cos\alpha + iy\sin\alpha} + \frac{d\alpha}{r}\right] = \frac{4\pi}{r} \qquad z > 0,$$

$$= 0 \qquad z < 0.$$

The electrostatic field derived from the function V is zero on one side of the plane $z = 0$ and has the character of the field due to a point charge on the other side. It should be noticed that the integrand is a potential function which becomes infinite along the line $z = 0$, $x\cos\alpha + y\sin\alpha = 0$, and as α varies this line sweeps out the plane of discontinuity of our electrostatic field.

To generalise this result we must endeavour to solve the

* *Aether and Matter* (1900), p. 188. The question as to whether the aether is continuous or discontinuous is discussed by H. Witte, *Ann. d. Phys.* (4), Bd. 26 (1908).

† Presidential Address, *British Association Reports*, Winnipeg (1909).

‡ *Recent Researches on Electricity and Magnetism*; *Electricity and Matter*; *Phil. Mag.* Vol. 19 (1910), p. 301, Oct.—Dec. (1912). See also N. R. Campbell, *The New Quarterly* (1909); *Phil. Mag.* Vol. 19 (1910), p. 181.

general problem of finding electromagnetic fields whose singularities lie on moving curves*.

A partial solution of this problem may be obtained by considering first of all the field represented by equations (10) of § 5. We may obtain a suitable pair of functions α, β by solving the equations

$$\left.\begin{aligned} \left(\frac{\partial\alpha}{\partial x}\right)^2 + \left(\frac{\partial\alpha}{\partial y}\right)^2 + \left(\frac{\partial\alpha}{\partial z}\right)^2 &= \frac{1}{c^2}\left(\frac{\partial\alpha}{\partial t}\right)^2 \\ \left(\frac{\partial\beta}{\partial x}\right)^2 + \left(\frac{\partial\beta}{\partial y}\right)^2 + \left(\frac{\partial\beta}{\partial z}\right)^2 &= \frac{1}{c^2}\left(\frac{\partial\beta}{\partial t}\right)^2 \\ \frac{\partial\alpha}{\partial x}\frac{\partial\beta}{\partial x} + \frac{\partial\alpha}{\partial y}\frac{\partial\beta}{\partial y} + \frac{\partial\alpha}{\partial z}\frac{\partial\beta}{\partial z} &= \frac{1}{c^2}\frac{\partial\alpha}{\partial t}\frac{\partial\beta}{\partial t} \end{aligned}\right\} \quad \ldots\ldots(279);$$

for clearly

$$\begin{aligned} &\left(\frac{\partial\alpha}{\partial y}\frac{\partial\beta}{\partial z} - \frac{\partial\alpha}{\partial z}\frac{\partial\beta}{\partial y}\right)^2 \\ &= \left[\left(\frac{\partial\alpha}{\partial y}\right)^2 + \left(\frac{\partial\alpha}{\partial z}\right)^2\right]\left[\left(\frac{\partial\beta}{\partial y}\right)^2 + \left(\frac{\partial\beta}{\partial z}\right)^2\right] - \left(\frac{\partial\alpha}{\partial y}\frac{\partial\beta}{\partial y} + \frac{\partial\alpha}{\partial z}\frac{\partial\beta}{\partial z}\right)^2 \\ &= \left[\frac{1}{c^2}\left(\frac{\partial\alpha}{\partial t}\right)^2 - \left(\frac{\partial\alpha}{\partial x}\right)^2\right]\left[\frac{1}{c^2}\left(\frac{\partial\beta}{\partial t}\right)^2 - \left(\frac{\partial\beta}{\partial x}\right)^2\right] - \left(\frac{1}{c^2}\frac{\partial\alpha}{\partial t}\frac{\partial\beta}{\partial t} - \frac{\partial\alpha}{\partial x}\frac{\partial\beta}{\partial x}\right)^2 \\ &= -\frac{1}{c^2}\left(\frac{\partial\alpha}{\partial x}\frac{\partial\beta}{\partial t} - \frac{\partial\alpha}{\partial t}\frac{\partial\beta}{\partial x}\right)^2. \end{aligned}$$

Two other equations can be obtained in a similar way and so it follows that if we make a suitable choice of an ambiguous sign which is involved in the definitions of the functions α and β, the equations (10) will be a consequence of equations (279).

A more general electromagnetic field is obtained by multiplying the components of M in equations (10) by an arbitrary function $f(\alpha, \beta)$. Since the components of M are necessarily solutions of the wave-equation it follows that, if $g = 1/f$, an expression of the type

$$\frac{\partial(\alpha, \beta)}{\partial(y, z)}\, g(\alpha, \beta)$$

is a solution of the wave-equation (8).

* The aim must be not only to obtain a complete generalisation of Green's equivalent layer which will be applicable to the case of a moving surface, but to obtain, if possible, an analysis of the electric and magnetic current sheets which are required. Cf. p. 29.

The electromagnetic field which has just been obtained generally has singularities at space-time points for which $f(\alpha, \beta)$ is zero. Let us write $f = u(x, y, z, t) + iv(x, y, z, t)$ where u and v are real when x, y, z, t are real; then the points for which $f = 0$ lie on the moving curve, defined by the equations $u = 0$, $v = 0$. Now it follows from (279) that f is a solution of the equation

$$\left(\frac{\partial f}{\partial x}\right)^2 + \left(\frac{\partial f}{\partial y}\right)^2 + \left(\frac{\partial f}{\partial z}\right)^2 = \frac{1}{c^2}\left(\frac{\partial f}{\partial t}\right)^2$$

and consequently

$$\left(\frac{\partial u}{\partial x}\right)^2 + \left(\frac{\partial u}{\partial y}\right)^2 + \left(\frac{\partial u}{\partial z}\right)^2 - \frac{1}{c^2}\left(\frac{\partial u}{\partial t}\right)^2 = \left(\frac{\partial v}{\partial x}\right)^2 + \left(\frac{\partial v}{\partial y}\right)^2 + \left(\frac{\partial v}{\partial z}\right)^2 - \frac{1}{c^2}\left(\frac{\partial v}{\partial t}\right)^2,$$

$$\frac{\partial u}{\partial x}\frac{\partial v}{\partial x} + \frac{\partial u}{\partial y}\frac{\partial v}{\partial y} + \frac{\partial u}{\partial z}\frac{\partial v}{\partial z} = \frac{1}{c^2}\frac{\partial u}{\partial t}\frac{\partial v}{\partial t}.$$

Now let $F(u, v) = 0$ be the equation of a moving surface which always contains the moving curve, then if

$$p = \frac{\partial F}{\partial u}\frac{\partial v}{\partial x} - \frac{\partial F}{\partial v}\frac{\partial u}{\partial x}, \qquad q = \frac{\partial F}{\partial u}\frac{\partial v}{\partial y} - \frac{\partial F}{\partial v}\frac{\partial u}{\partial y},$$

$$r = \frac{\partial F}{\partial u}\frac{\partial v}{\partial z} - \frac{\partial F}{\partial v}\frac{\partial u}{\partial z}, \qquad s = \frac{\partial F}{\partial u}\frac{\partial v}{\partial t} - \frac{\partial F}{\partial v}\frac{\partial u}{\partial t},$$

we have

$$\left[\left(\frac{\partial F}{\partial x}\right)^2 + \left(\frac{\partial F}{\partial y}\right)^2 + \left(\frac{\partial F}{\partial z}\right)^2 - \frac{1}{c^2}\left(\frac{\partial F}{\partial t}\right)^2\right]$$

$$\times\left[\left(\frac{\partial F}{\partial x}\right)^2 + \left(\frac{\partial F}{\partial y}\right)^2 + \left(\frac{\partial F}{\partial z}\right)^2 + \frac{s^2}{c^2}\right]$$

$$= \left(q\frac{\partial F}{\partial z} - r\frac{\partial F}{\partial y}\right)^2 + \left(r\frac{\partial F}{\partial x} - p\frac{\partial F}{\partial z}\right)^2 + \left(p\frac{\partial F}{\partial y} - q\frac{\partial F}{\partial x}\right)^2;$$

consequently

$$\left(\frac{\partial F}{\partial x}\right)^2 + \left(\frac{\partial F}{\partial y}\right)^2 + \left(\frac{\partial F}{\partial z}\right)^2 \nless \frac{1}{c^2}\left(\frac{\partial F}{\partial t}\right)^2.$$

This means that the component velocity of the surface $F = 0$ in the direction of the normal at (x, y, z, t) is less than the velocity of light, it is equal to the velocity of light only in an exceptional case. Since this is true for any surface that always contains the moving curve it follows that the curve can be regarded as moving with a velocity less than that of light. It should be understood that the curve generally changes in shape as it moves but this is not necessarily the case.

It often happens that the moving curve $f=0$ can be regarded as made up of the instantaneous positions of a series of points which move in straight lines with the velocity of light. To see this let us suppose that a function $\phi(\alpha, \beta)$ can be found such that ϕ is a *real* function of x, y, z, t. It is evident from (279) that

$$\frac{\partial f}{\partial x}\frac{\partial \phi}{\partial x}+\frac{\partial f}{\partial y}\frac{\partial \phi}{\partial y}+\frac{\partial f}{\partial z}\frac{\partial \phi}{\partial z}=\frac{1}{c^2}\frac{\partial f}{\partial t}\frac{\partial \phi}{\partial t},$$

and this means that if a point starts from (x, y, z, t) and moves with the velocity of light along a straight line whose direction cosines l, m, n are proportional to $\frac{\partial \phi}{\partial x}, \frac{\partial \phi}{\partial y}, \frac{\partial \phi}{\partial z}$, the function f will remain constant along its path, and consequently if the point once lies on the moving curve $f=0$ it will always lie on this curve. It should be noticed that the function ϕ and its first derivatives with regard to x, y, z, t all remain constant along the path of the moving point.

The case in which no such function ϕ exists may be of importance in future developments of the subject; this case has not yet been discussed in any detail.

Two methods of solving equations (279) are known, but they are not really distinct. In the first method the functions α, β are defined by equations

$$\left.\begin{aligned}[x-\xi(\alpha,\beta)]^2+[y-\eta(\alpha,\beta)]^2+[z-\zeta(\alpha,\beta)]^2=c^2[t-\tau(\alpha,\beta)]^2\\ l(\alpha,\beta)[x-\xi(\alpha,\beta)]+m(\alpha,\beta)[y-\eta(\alpha,\beta)]\\ +n(\alpha,\beta)[z-\zeta(\alpha,\beta)]=c^2p(\alpha,\beta)[t-\tau(\alpha,\beta)]\end{aligned}\right\} \quad \text{......(280)},$$

where $\xi, \eta, \zeta, \tau, l, m, n, p$ are arbitrary functions satisfying the relation

$$l^2+m^2+n^2=c^2p^2.$$

The functions α, β may evidently be replaced by two other functions α', β' defined by equations such as

$$\alpha'=F(\alpha,\beta), \qquad \beta'=G(\alpha,\beta);$$

consequently we may without loss of generality introduce a

further relation connecting the functions $\xi, \eta, \zeta, \tau, l, m, n, p$. The relation we shall choose is

$$l\frac{\partial \xi}{\partial \beta} + m\frac{\partial \eta}{\partial \beta} + n\frac{\partial \zeta}{\partial \beta} = c^2 p\frac{\partial \tau}{\partial \beta} \quad \text{..........(281).}$$

When α and β are defined in this way we can obtain the simple specifications of two types of electromagnetic fields; one type is obtained by writing

$$f(\alpha, \beta) M_x = \frac{\partial(\alpha, \beta)}{\partial(y, z)} = \pm \frac{i}{c}\frac{\partial(\alpha, \beta)}{\partial(x, t)} \quad \text{........(282),}$$

and two similar equations. A second type of electromagnetic field may be derived from the potentials*

$$\left.\begin{aligned} A_x &= \frac{lSg(\alpha, \beta)}{PS - QR}, \quad A_y = \frac{mSg(\alpha, \beta)}{PS - QR} \\ A_z &= \frac{nSg(\alpha, \beta)}{PS - QR}, \quad \Phi = \frac{cpSg(\alpha, \beta)}{PS - QR} \end{aligned}\right\} \quad \text{......(283),}$$

where

$$P = \frac{\partial \xi}{\partial \alpha}(x - \xi) + \frac{\partial \eta}{\partial \alpha}(y - \eta) + \frac{\partial \zeta}{\partial \alpha}(z - \zeta) - c^2\frac{\partial \tau}{\partial \alpha}(t - \tau),$$

$$Q = \frac{\partial \xi}{\partial \beta}(x - \xi) + \frac{\partial \eta}{\partial \beta}(y - \eta) + \frac{\partial \zeta}{\partial \beta}(z - \zeta) - c^2\frac{\partial \tau}{\partial \beta}(t - \tau),$$

$$\begin{aligned} R = l\frac{\partial \xi}{\partial \alpha} + m\frac{\partial \eta}{\partial \alpha} + n\frac{\partial \zeta}{\partial \alpha} - c^2 p\frac{\partial \tau}{\partial \alpha} - \frac{\partial l}{\partial \alpha}(x - \xi) \\ - \frac{\partial m}{\partial \alpha}(y - \eta) - \frac{\partial n}{\partial \alpha}(z - \zeta) + c^2\frac{\partial p}{\partial \alpha}(t - \tau), \end{aligned}$$

$$S = c^2\frac{\partial p}{\partial \beta}(t - \tau) - \frac{\partial l}{\partial \beta}(x - \xi) - \frac{\partial m}{\partial \beta}(y - \eta) - \frac{\partial n}{\partial \beta}(z - \zeta).$$

It is easy to see that these potentials satisfy the relation (267). To prove that they are wave-functions we remark that

$$\begin{aligned} A_x = \frac{g(\alpha, \beta)}{PS - QR}\Big[\frac{\partial m}{\partial \beta}\{m(x - \xi) - l(y - \eta)\} \\ + \frac{\partial n}{\partial \beta}\{n(x - \xi) - l(z - \zeta)\} - c^2\frac{\partial p}{\partial \beta}\{p(x - \xi) - l(t - \tau)\}\Big] \\ = g\left[\frac{\partial m}{\partial \beta}\cdot\frac{\partial(\alpha, \beta)}{\partial(x, y)} + \frac{\partial n}{\partial \beta}\cdot\frac{\partial(\alpha, \beta)}{\partial(x, z)} + \frac{\partial p}{\partial \beta}\cdot\frac{\partial(\alpha, \beta)}{\partial(x, t)}\right]. \end{aligned}$$

Now it has already been proved that an arbitrary function of α and β multiplied by one of the Jacobians represents a

* We can also call these potentials L_x, L_y, L_z, Λ and derive an electromagnetic field from them by the method of § 4.

wave-function, hence A_x is a wave-function. In a similar way it can be shown that the other potentials are wave-functions.

It should be noticed that the electromagnetic field specified by the potentials (283) is conjugate to the field given by (282). To prove this we observe that

$$A_x = \frac{gS\left(R\dfrac{\partial \alpha}{\partial x} + S\dfrac{\partial \beta}{\partial x}\right)}{PS - QR} = u\frac{\partial \alpha}{\partial x} + v\frac{\partial \beta}{\partial x} \text{ say,}$$

and there are similar expressions for A_y, A_z, Φ. Hence for the electromagnetic field specified by the potentials (283), we have

$$M_x' = \frac{\partial (u, \alpha)}{\partial (y, z)} + \frac{\partial (v, \beta)}{\partial (y, z)} \pm \frac{i}{c}\frac{\partial (u, \alpha)}{\partial (x, t)} \pm \frac{i}{c}\frac{\partial (v, \beta)}{\partial (x, t)}.$$

The relation (3) is now seen to be satisfied in virtue of two equations of type

$$\frac{\partial (\alpha, \beta)}{\partial (y, z)}\frac{\partial (u, \alpha)}{\partial (x, t)} + \frac{\partial (\alpha, \beta)}{\partial (z, x)}\frac{\partial (u, \alpha)}{\partial (y, t)} + \frac{\partial (\alpha, \beta)}{\partial (x, y)}\frac{\partial (u, \alpha)}{\partial (z, t)}$$
$$+ \frac{\partial (\alpha, \beta)}{\partial (x, t)}\frac{\partial (u, \alpha)}{\partial (y, z)} + \frac{\partial (\alpha, \beta)}{\partial (y, t)}\frac{\partial (u, \alpha)}{\partial (z, x)} + \frac{\partial (\alpha, \beta)}{\partial (z, t)}\frac{\partial (u, \alpha)}{\partial (x, y)} = 0 \quad \text{......(284).}$$

It should be remarked that the vectors E, H in both fields generally become infinite when $PS - QR = 0$. This equation is certainly satisfied when (x, y, z, t) lies on the moving curve defined by the equations

$$x = \xi(\alpha, \beta), \quad y = \eta(\alpha, \beta), \quad z = \zeta(\alpha, \beta), \quad t = \tau(\alpha, \beta) \text{...(285).}$$

In some cases this curve may reduce to a moving point, as for instance when ξ, η, ζ, τ are independent of β.

It is evident that the quantity $PS - QR$ is usually zero for space-time points which do not lie on the moving curve (285). If, however, we regard l, m, n, p as complex functions of the type $\phi(\alpha, \beta) + i\psi(\alpha, \beta)$, the equation $PS = QR$ will generally give rise to two distinct equations connecting x, y, z, t, when we equate real and imaginary terms on both sides. Hence all the real singularities that are defined by $PS = QR$ will generally lie on one or a number of moving curves.

So far we have said nothing about the choice of a suitable pair of roots of equations (280). In general we cannot expect them both to be real and it is difficult to lay down rules which will enable us to pick out just one α and just one β in all cases. To proceed further we must consider some particular examples; before we do this, however, it will be worth while to point out that if we assign given complex values to α and β the equations (280) will generally determine two real points x, y, z, t, but in special cases they may give ∞^1 space-time points which can be regarded as the consecutive positions of a moving point. Thus we have an interesting specification of the real points in space by means of two complex quantities. If we assign a complex value to β the corresponding space-time points x, y, z, t generally lie on a moving curve which travels with a velocity not greater than that of light. Hence in the parametric representation of x, y, z, t in terms of the complex quantities α, β the loci $\alpha = \text{const.}$, $\beta = \text{const.}$ are generally moving curves which may alter in shape as they move but never travel with a velocity greater than that of light.

Matters are somewhat different if α or some function of α and β is always real when x, y, z, t are real. This case will now be illustrated by a particular example.

Examples. 1. Prove that the ratios of the Jacobians

$$\frac{\partial(\alpha, \beta)}{\partial(y, z)}, \quad \frac{\partial(\alpha, \beta)}{\partial(z, x)}, \quad \frac{\partial(\alpha, \beta)}{\partial(x, y)}$$

are functions of α and β.

2. Prove that the ratio Q/S depends only on α and β.

3. Obtain the general solution of equations (279) by taking x, y, α, β as new independent variables.

§ 44. Projection of singularities from a moving point, second model of a corpuscle.

Let us now suppose that ξ, η, ζ, τ are independent of β and that $\tau = \alpha$. We may then define α uniquely by restricting it to be real and introducing the inequalities

$$\left(\frac{\partial \xi}{\partial \alpha}\right)^2 + \left(\frac{\partial \eta}{\partial \alpha}\right)^2 + \left(\frac{\partial \zeta}{\partial \alpha}\right)^2 < c^2, \quad \alpha \leqslant t \quad \text{.........(286).}$$

To obtain a single value of β we may assume l, m, n, p to be linear functions of β. Consequently we may put

$$\beta = \frac{l_0(x-\xi) + m_0(y-\eta) + n_0(z-\zeta) - c^2 p_0 (t-\alpha)}{l_1(x-\xi) + m_1(y-\eta) + n_1(z-\zeta) - c^2 p_1 (t-\alpha)} \text{...(287),}$$

where l_0, l_1, etc. are functions of α which satisfy the equations

$$\left.\begin{aligned} l_0^2 + m_0^2 + n_0^2 &= c^2 p_0^2 \\ l_1^2 + m_1^2 + n_1^2 &= c^2 p_1^2 \\ l_0 l_1 + m_0 m_1 + n_0 n_1 &= c^2 p_0 p_1 \end{aligned}\right\} \quad \ldots\ldots\ldots\ldots(288).$$

It is easy to verify that the wave-equation is satisfied by a function of type

$$\Omega = \frac{1}{P} g(\alpha, \beta);$$

thus we have a generalisation of the theorem of § 41.

We shall assume that l_0, m_0, n_0, p_0 are real and that some or all of the quantities l_1, m_1, n_1, p_1 are complex. It is easy to see that if we assign a real value to α and a complex value to β the corresponding space-time points (x, y, z, t) can be regarded as the successive positions of a point which starts from the point

$$x = \xi(\alpha), \quad y = \eta(\alpha), \quad z = \zeta(\alpha), \quad t = \alpha \ldots\ldots(289)$$

and moves with the velocity of light along a straight line through this point. There is clearly just one line through this point for each complex value of β and *vice versa*. If we consider all the points in space at a particular time t we can specify each point uniquely by a real parameter α and a real or complex parameter β.

Let us now consider the electromagnetic field which is specified by the potentials

$$A_x = \frac{lf}{P}, \quad A_y = \frac{mf}{P}, \quad A_z = \frac{nf}{P}, \quad \Phi = \frac{cpf}{P} \ldots\ldots(290),$$

where $l = \beta l_1 - l_0$, $m = \beta m_1 - m_0$, $n = \beta n_1 - n_0$, $p = \beta p_1 - p_0$ and f is an arbitrary function of α and β. These potentials are derived from (283) by putting $Q = 0$.

After a long calculation we find that the component of the electric force along the radius from $(\xi, \eta, \zeta, \alpha)$ to (x, y, z, t) is

$$-\frac{f}{P^2}\left[c^2 p - l\frac{\partial \xi}{\partial \alpha} - m\frac{\partial \eta}{\partial \alpha} - n\frac{\partial \zeta}{\partial \alpha}\right].$$

To obtain an electromagnetic field in which there is a constant electric charge associated with the singularity $(\xi, \eta, \zeta, \alpha)$, we assume that $p_0 = p_1 = f = 1$ and that

$$\left.\begin{aligned} l_1\frac{\partial\xi}{\partial\alpha}+m_1\frac{\partial\eta}{\partial\alpha}+n_1\frac{\partial\zeta}{\partial\alpha}&=c^2\\ l_0\frac{\partial\xi}{\partial\alpha}+m_0\frac{\partial\eta}{\partial\alpha}+n_0\frac{\partial\zeta}{\partial\alpha}&=\left(\frac{\partial\xi}{\partial\alpha}\right)^2+\left(\frac{\partial\eta}{\partial\alpha}\right)^2+\left(\frac{\partial\zeta}{\partial\alpha}\right)^2\end{aligned}\right\}\ldots(291).$$

The expression for the radial electric force then becomes

$$\frac{1}{P^2}\left[c^2-\xi'^2-\eta'^2-\zeta'^2\right].$$

Comparing this with the expression for the radial electric force in the case of an electromagnetic field with a simple singularity (ξ, η, ζ, τ), we see that there is a constant electric charge $4\pi/c$ associated with the singularity $(\xi, \eta, \zeta, \alpha)$.

It should be mentioned that the second of equations (291) is a consequence of the other equations satisfied by l_0, m_0, n_0. To prove this we take the axis of x in a direction parallel to the velocity of the singularity (ξ, η, ζ) at time α. We then have for this instant $\frac{\partial\eta}{\partial\alpha}=\frac{\partial\zeta}{\partial\alpha}=0$. If, moreover, we choose the axis of y in such a way that $n_1=0$, we may satisfy the first of equations (291) by writing

$$\frac{\partial\xi}{\partial\alpha}=c\cos\theta,\quad l_1=c\sec\theta,\quad m_1=ic\tan\theta,\quad l_0=c\cos\theta,$$
$$m_0=0,\quad n_0=\pm c\sin\theta,$$

and then it is clear that the second of equations (291) is satisfied.

Let us now write $\bar{l}_0=c\cos\theta$, $\bar{m}_0=0$, $\bar{n}_0=\mp c\sin\theta$, $\bar{l}_1=c\sec\theta$, $\bar{m}_1=-ic\tan\theta$, $\bar{n}_1=0$, $x-\xi=X$, $y-\eta=Y$, $z-\zeta=Z$, $t-\tau=T$; then it is easy to see that if

$$S=c^2T-l_1X-m_1Y-n_1Z,\qquad \bar{S}=c^2T-\bar{l}_1X-\bar{m}_1Y-\bar{n}_1Z,$$
$$U=c^2T-l_0X-m_0Y-n_0Z,\qquad \bar{U}=c^2T-\bar{l}_0X-\bar{m}_0Y-\bar{n}_0Z,$$

we have $\qquad \cos^2\theta\, S\bar{S}=U\bar{U},\qquad U+\bar{U}=-2P.$

Now since $\beta S=U$ it follows that the potentials (290) become infinite when $\bar{U}=0$, i.e. when

$$\frac{X}{\bar{l}_0}=\frac{Y}{\bar{m}_0}=\frac{Z}{\bar{n}_0}=\frac{T}{1}.$$

When α is given these equations are satisfied by a point which starts at $(\xi, \eta, \zeta, \alpha)$ and moves with the velocity of light along a straight line whose direction cosines are proportional to $\bar{l}_0, \bar{m}_0, \bar{n}_0$. This line makes an angle θ with the direction of motion of the point $(\xi, \eta, \zeta, \alpha)$.

The electric and magnetic forces in the electromagnetic field derived from the potentials (290) are generally infinite for $\bar{U}=0$ and so our field possesses a number of singular points which are projected from the moving point $(\xi, \eta, \zeta, \alpha)$ and travel along straight lines with the velocity of light. It should be remarked, however, that if we retain only the real parts of the potentials (290), the projected singularities disappear as soon as the singularity ξ, η, ζ, α moves in a straight line with uniform velocity and l, m, n are independent of α. The field then becomes identical with that derived from Liénard's potentials. To prove this we shall show that, on the above assumptions, the field derived from the potentials

$$A_x^0 = \mathbf{R}\,\frac{l-\xi'}{P}, \quad A_y^0 = \mathbf{R}\,\frac{m-\eta'}{P},$$

$$A_z^0 = \mathbf{R}\,\frac{n-\zeta'}{P}, \quad \Phi^0 = \mathbf{R}\,\frac{cp-1}{P} \quad \text{.........(292)},$$

is everywhere null, $\mathbf{R}$ being used to denote the real part of a quantity following it. In the first place we remark that we now have

$$R = l\xi' + m\eta' + n\zeta' - c^2 p = \xi'^2 + \eta'^2 + \zeta'^2 - c^2,$$

$$\frac{\partial}{\partial x}(\log P) = \frac{\xi'}{P} + \frac{1}{P}\frac{\partial \alpha}{\partial x}(c^2 - \xi'^2 - \eta'^2 - \zeta'^2).$$

Hence
$$2A_x^0 = 2\mathbf{R}\left[\frac{1}{P}\left(R\frac{\partial \alpha}{\partial x} + S\frac{\partial \beta}{\partial x} - \xi'\right)\right],$$

$$= \frac{S}{P}\frac{\partial}{\partial x}\left(\frac{U}{S}\right) + \frac{\bar{S}}{P}\frac{\partial}{\partial x}\left(\frac{U}{\bar{S}}\right) - 2\frac{\partial}{\partial x}(\log P),$$

$$= -\frac{U}{P}\frac{\partial}{\partial x}\log\frac{S\bar{S}}{U^2} - 2\frac{\partial}{\partial x}(\log P),$$

$$= -\frac{U}{P}\frac{\partial}{\partial x}\log\frac{\bar{U}}{U} - 2\frac{\partial}{\partial x}(\log P).$$

Now since $U + \bar{U} + 2P = 0$, it follows that U/P is a function c $\bar{U}/U$ and so $A_x^0, A_y^0, A_z^0, -\frac{1}{c}\Phi^0$ are the derivatives of a singl function, consequently the electromagnetic field derived from these potentials is everywhere null.

Summing up our results we can say that when the conditions (291) are satisfied, the electromagnetic field derived from the potentials (290) contains a point charge which moves with a velocity less than that of light; attached to this point charge there is a certain curve which becomes the locus of a series of moving point singularities whenever its form differs in any portion from a straight line or its direction changes. The form of the curve at any instant is subject to the condition that the points of the curve can be regarded as having been projected from the moving charge at different instants, the direction of projection being partially determined by the law $\cos\theta = v/c$ where v is the velocity of the point charge, and θ is the angle between the direction of projection and the direction of motion of the point charge.

We may now obtain a new model of a corpuscle by considering an aggregate of elementary fields of the type just described, the point charges and exceptional curves being nearly coincident. If we write de for the charge associated with one of the elementary fields we may obtain a field in which the electric and magnetic forces are finite by a suitable process of integration. According to this idea a corpuscle has a kind of tube or thread attached to it. When the motion of the corpuscle changes a wave or kink runs along the thread; the energy radiated from the corpuscle spreads out in all directions but is concentrated round the thread so that the thread acts as a guiding wire. This theory of radiation is in some respects similar to that given by Sir Joseph Thomson in his theory of the Röntgen rays*. It is in accordance with his idea that the energy may be concentrated round certain points of the wave-front.

The following figure indicates roughly the changes in the form of a tube which always lies in one plane and is attached to a corpuscle performing a simple harmonic motion; it is seen that a type of progressive wave travels along the tube.

* *Electricity and Matter*, London (1904); *Phil. Mag.* Vol. 19 (1910).

In Sir Joseph Thomson's theory of the Röntgen rays the kink in the tube of force becomes longer and longer as it recedes from the charge. A similar remark applies to the

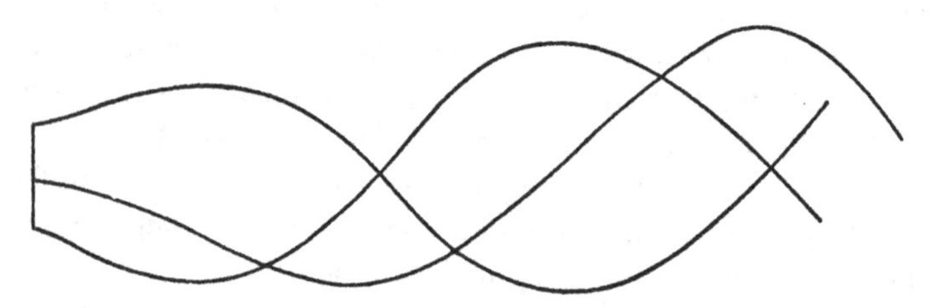

oscillations of the thread attached to our point charge. This phenomenon may be due entirely to the fact that the tube of force and thread extend to infinity. If we suppose that the tube or thread does not extend to infinity but ends at some other point charge, the circumstances of the motion will be different. If in this case we treat the thread as a singular line of an electromagnetic field and suppose that it is given by an equation of the form

$$f(\alpha, \beta) = 0$$

where α and β are functions which satisfy (279), we must conclude that there is no function of type $F(\alpha, \beta)$ which is a real function of x, y, z, t; for if this were the case the moving thread would be the locus of points travelling in straight lines with the velocity of light and would consequently extend to infinity.

Electromagnetic fields with moving point charges joined by singular curves which do not extend to infinity have not yet been obtained, but I think there is some hope of deriving them by the general methods of § 43, when the quantities α, β are both complex.

Examples. 1. Discuss the properties of the electromagnetic fields that can be derived from the potentials

$$A_x = \frac{l_1}{s} + \frac{\bar{l}_1}{\bar{s}}, \quad A_y = \frac{m_1}{s} + \frac{\bar{m}_1}{\bar{s}}, \quad A_z = \frac{n_1}{s} + \frac{\bar{n}_1}{\bar{s}}, \quad \Phi = \frac{cp_1}{s} + \frac{c\bar{p}_1}{\bar{s}};$$

$$A_x' = \frac{l_1}{is} - \frac{\bar{l}_1}{i\bar{s}}, \quad A_y' = \frac{m_1}{is} - \frac{\bar{m}_1}{i\bar{s}}, \quad A_z' = \frac{n_1}{is} - \frac{\bar{n}_1}{i\bar{s}}, \quad \Phi' = \frac{cp_1}{is} - \frac{c\bar{p}_1}{i\bar{s}};$$

respectively and determine the lines of electric and magnetic force.

2. Prove that if α and β are defined as in § 44, the electromagnetic field specified by the equations (282) is conjugate to the field specified by the potentials (290), or to the field specified by Liénard's potentials.

§ 45. Electromagnetic fields with singularities moving with velocities greater than that of light.

Some of the preceding analysis holds and provides us with solutions of Maxwell's equations when the velocity of the primary singularity (ξ, η, ζ, τ) is greater than that of light, but in the case of a field specified by potentials of type (266) a transition from a velocity less than that of light to a velocity greater than that of light does not seem to be physically possible on account of the occurrence of infinite values of the electric and magnetic forces in the critical case. Moreover, it is difficult in the general case to give a rule which will enable us to pick out just one root of the equation (265). An interesting type of field may, however, be obtained by a process of summation over some of the roots of the equation *.

The case of infinite velocity is of some interest, for then we obtain electromagnetic fields with singularities along a fixed curve at a given instant of time. The following example indicates that the case in which the primary singularities are imaginary may be associated with another case in which they are real.

Consider the two equations

$$(x - a\cos\alpha)^2 + (y - a\sin\alpha)^2 + z^2 = c^2t^2,$$
$$x^2 + y^2 + (z - ai\cosh\beta)^2 = (ct - ia\sinh\beta)^2,$$

and write in analogy with (263)

$$\nu = a\cos\alpha\,(y - a\sin\alpha) - a\sin\alpha\,(x - a\cos\alpha) = a\,(y\cos\alpha - x\sin\alpha),$$
$$\nu_1 = ia\sinh\beta\,(z - ia\cosh\beta) - ia\cosh\beta\,(ct - ia\sinh\beta)$$
$$= ia\,(z\sinh\beta - ct\cosh\beta).$$

We evidently have

$$4\nu^2 = 4a^2(x^2 + y^2) - (x^2 + y^2 + z^2 - c^2t^2 + a^2)^2,$$
$$4\nu_1^2 = 4a^2(z^2 - c^2t^2) + (x^2 + y^2 + z^2 - c^2t^2 - a^2)^2,$$

hence
$$\nu^2 + \nu_1^2 = 0.$$

* A more complete discussion is given in G. A. Schott's *Electromagnetic Radiation.*

It is easy to verify that the function

$$u = \frac{1}{\nu} f(\alpha, \beta)$$

is a wave-function and that the functions α, β are suitable for obtaining electromagnetic fields by the method of § 5; they are, in fact, the functions considered in § 36.

§ 46. Second solution of the fundamental equations.

The fundamental equations (279) are also satisfied when the functions (α, β) are defined by the relations

$$\left.\begin{aligned} l(\alpha,\beta)x + m(\alpha,\beta)y + n(\alpha,\beta)z - c^2 p(\alpha,\beta)t + g(\alpha,\beta) = 0 \\ L(\alpha,\beta)x + M(\alpha,\beta)y + N(\alpha,\beta)z - c^2 P(\alpha,\beta)t + G(\alpha,\beta) = 0 \end{aligned}\right\} \quad \text{.........(293)}$$

where l, L, etc. are arbitrary functions satisfying the relations

$$l^2 + m^2 + n^2 = c^2p^2, \qquad L^2 + M^2 + N^2 = c^2P^2,$$

$$lL + mM + nN = c^2pP \quad \text{..............(294)}.$$

If now

$$\lambda = x\frac{\partial l}{\partial \alpha} + y\frac{\partial m}{\partial \alpha} + z\frac{\partial n}{\partial \alpha} - c^2t\frac{\partial p}{\partial \alpha} + \frac{\partial g}{\partial \alpha},$$

$$\mu = x\frac{\partial l}{\partial \beta} + y\frac{\partial m}{\partial \beta} + z\frac{\partial n}{\partial \beta} - c^2t\frac{\partial p}{\partial \beta} + \frac{\partial g}{\partial \beta},$$

$$\nu = x\frac{\partial L}{\partial \alpha} + y\frac{\partial M}{\partial \alpha} + z\frac{\partial N}{\partial \alpha} - c^2t\frac{\partial P}{\partial \alpha} + \frac{\partial G}{\partial \alpha},$$

$$\varpi = x\frac{\partial L}{\partial \beta} + y\frac{\partial M}{\partial \beta} + z\frac{\partial N}{\partial \beta} - c^2t\frac{\partial P}{\partial \beta} + \frac{\partial G}{\partial \beta};$$

we can derive an electromagnetic field from the potentials

$$A_x = \frac{L\sigma f}{\lambda\varpi - \mu\nu}, \quad A_y = \frac{M\sigma f}{\lambda\varpi - \mu\nu}, \quad A_z = \frac{N\sigma f}{\lambda\varpi - \mu\nu}, \quad \Phi = \frac{cP\sigma f}{\lambda\varpi - \mu\nu} \quad \text{.........(295)}$$

where f is an arbitrary function of α and β, and

$$\sigma = l\frac{\partial L}{\partial \beta} + m\frac{\partial M}{\partial \beta} + n\frac{\partial N}{\partial \beta} - c^2p\frac{\partial P}{\partial \beta} \quad \text{.........(296)}.$$

To verify that these potentials satisfy the wave-equation and the relation (267), it is sufficient to remark that

$$\left.\begin{aligned} f^{-1} A_x &= -\frac{\partial M}{\partial \beta}\frac{\partial(\alpha,\beta)}{\partial(x,y)} - \frac{\partial N}{\partial \beta}\frac{\partial(\alpha,\beta)}{\partial(x,z)} - \frac{\partial P}{\partial \beta}\frac{\partial(\alpha,\beta)}{\partial(x,t)} \\ &= \mp\frac{i}{c}\frac{\partial M}{\partial \beta}\frac{\partial(\alpha,\beta)}{\partial(z,t)} \pm \frac{i}{c}\frac{\partial N}{\partial \beta}\frac{\partial(\alpha,\beta)}{\partial(y,t)} \pm ic\frac{\partial P}{\partial \beta}\frac{\partial(\alpha,\beta)}{\partial(y,z)} \end{aligned}\right\} \ldots (297).$$

When similar expressions are obtained for A_y, A_z, Φ it is clear that the relation (267) is satisfied. Other types of electromagnetic fields are obtained by writing α instead of β in equation (296) or by writing l, m, n, p in place of L, M, N, P in (295) and (296). If A_x', A_y', A_z', Φ' are the potentials obtained in either of these ways we have clearly

$$A_x A_x' + A_y A_y' + A_z A_z' - \Phi\Phi' = 0 \ldots\ldots\ldots (298).$$

I have noticed that this relation is often satisfied by the potentials of two conjugate fields.

Another type of electromagnetic field may be obtained by the method of § 5.

Some particular cases of the preceding theorems may be deduced by contour integration. To illustrate the method let us suppose that $l, m, \ldots L, M, \ldots$ are functions of a parameter α which satisfy the relations (294). If we regard these quantities as independent of x, y, z, t, the contour integral

$$V = \frac{1}{2\pi i}\int \frac{F(Lx + My + Nz - c^2Pt + G, \alpha)}{lx + my + nz - c^2pt + g}\, d\alpha$$

will represent a wave-function. Now let us suppose that the contour encloses only one root of the equation

$$xl(\alpha) + ym(\alpha) + zn(\alpha) - c^2tp(\alpha) + g(\alpha) = 0 \ldots (299),$$

and that the numerator is finite and single-valued within the contour and on its boundary, then by Cauchy's theorem the value of the integral is generally

$$\frac{1}{\lambda} F(\beta, \alpha) \ldots\ldots\ldots\ldots\ldots\ldots\ldots\ldots (300)$$

where $\beta = xL(\alpha) + yM(\alpha) + zN(\alpha) - c^2tP(\alpha) + G(\alpha) \ldots (301)$

and λ is defined in the same way as before.

We have then the result that the function (300) satisfies the wave-equation. This is a particular case of the general theorem of § 43 and is a generalisation of a theorem due to Forsyth*. In Forsyth's work the functions L, M, N, P are assumed to be the derivatives of l, m, n, p with regard to α. The functions α, β are evidently particular cases of the functions that have already been defined and so may be used to construct an electromagnetic field by the method of § 5. An interesting electromagnetic field may also be derived from the potentials

$$A_x = \frac{l'(\alpha)}{\lambda}, \quad A_y = \frac{m'(\alpha)}{\lambda}, \quad A_z = \frac{n'(\alpha)}{\lambda}, \quad \Phi = \frac{cp'(\alpha)}{\lambda} \quad \ldots(302);$$

it is easy to verify that the relation (267) is satisfied. The case in which l, m, n, p, g are all real functions of α is uninteresting because then our potentials become infinite for ∞^3 space-time points which lie in ∞^1 planes. When, however, l, m, n, p, g are complex functions of the type $\phi(\alpha) + i\psi(\alpha)$, the singularities of the electromagnetic field generally lie on a moving curve.

It should be remarked that when α and β are defined by the equations (299) and (301) a function of the type

$$\Omega = F(\alpha, \beta)$$

satisfies the wave-equation, and at the same time satisfies the differential equation

$$\left(\frac{\partial \Omega}{\partial x}\right)^2 + \left(\frac{\partial \Omega}{\partial y}\right)^2 + \left(\frac{\partial \Omega}{\partial z}\right)^2 = \frac{1}{c^2}\left(\frac{\partial \Omega}{\partial t}\right)^2.$$

This is a generalisation of a theorem due to Forsyth† and Jacobi‡. A solution of Maxwell's equations may be derived from the potentials

$$A_x^0 = l_0(\alpha), \quad A_y^0 = m_0(\alpha), \quad A_z^0 = n_0(\alpha), \quad \Phi^0 = cp_0(\alpha) \quad \ldots(303)$$

if

$$l\frac{\partial l_0}{\partial \alpha} + m\frac{\partial m_0}{\partial \alpha} + n\frac{\partial n_0}{\partial \alpha} - c^2 p\frac{\partial p_0}{\partial \alpha} = 0,$$

$$l_0\frac{\partial l}{\partial \alpha} + m_0\frac{\partial m}{\partial \alpha} + n_0\frac{\partial n}{\partial \alpha} - c^2 p_0\frac{\partial p}{\partial \alpha} = 0.$$

* *Messenger of Mathematics*, Vol. 27 (1898), p. 138. The theorem is obtained in another manner by Forsyth.

† *Loc. cit.*

‡ *Werke*, Bd. 2, p. 208.

One way of satisfying these equations is to put

$$l_0 = l, \quad m_0 = m, \quad n_0 = n, \quad p_0 = p.$$

An interesting type of wave-function may be obtained by a generalisation of a method due to Schottky *.

Let

$$a_0(\alpha) = \int l(\alpha)\, d\alpha, \quad b_0(\alpha) = \int m(\alpha)\, d\alpha, \quad c_0(\alpha) = \int n(\alpha)\, d\alpha,$$
$$d_0(\alpha) = \int p(\alpha)\, d\alpha, \quad e_0(\alpha) = \int g(\alpha)\, d\alpha,$$

then $\Omega = xa_0(\alpha) + yb_0(\alpha) + zc_0(\alpha) - c^2 t d_0(\alpha) + e_0(\alpha)$

is a wave-function whose derivatives with regard to x, y, z, t are all functions of the single variable α. The function α is supposed to be defined by equation (299).

Example. If $l = 1 - s\alpha$, $m = s + \alpha$, $n = i(s - \alpha)$, $cp = 1 + s\alpha$, $g = \alpha^2$, where s is a constant, the function λ vanishes when (x, y, z, t) lies on the moving curve

$$x = \frac{1 - s^2}{1 + s^2} ct + \frac{3s^2 - v^2}{1 + v^2}, \quad y = \frac{2s}{1 + s^2} ct + \frac{s(s^2 - v^2) - 2s}{1 + v^2}, \quad z = 2v.$$

A point for which v is constant moves in a straight line with the velocity of light.

§ 47. A wave-function with a fixed curve of singularities.

Let τ be defined in terms of x, y, z by the equation

$$a\tau = xp(\tau) + yq(\tau) + zr(\tau)$$

where $p^2 + q^2 + r^2 = 0$ and a is a constant.

Let $\qquad \nu = a - xp'(\tau) - yq'(\tau) - zr'(\tau),$

$$\theta = ct \pm \frac{xp' + yq' + zr'}{(p'^2 + q'^2 + r'^2)^{\frac{1}{2}}},$$

then $\Omega = \frac{1}{\nu} f(\theta, \tau)$ is a wave-function†.

A solution of Maxwell's equations may be derived from the potentials

$$A_x = \frac{p}{\nu} f(\theta, \tau), \quad A_y = \frac{q}{\nu} f(\theta, \tau), \quad A_z = \frac{r}{\nu} f(\theta, \tau), \quad \Phi = 0.$$

* *Berlin. Sitzungsberichte* (1909), p. 1152.

† This result was derived from a theorem given by Prof. Forsyth, *loc. cit.* If we put $\theta = \beta$, $\tau = \alpha$ the functions α, β can be used to obtain an electromagnetic field by the method of § 5. These functions are, of course, particular cases of the functions defined at the beginning of § 46.

To obtain a real electromagnetic field we must retain only the real parts of these expressions.

It can be shown by putting

$$p = P_1(u, v) + iP_2(u, v), \qquad q = Q_1(u, v) + iQ_2(u, v),$$
$$r = R_1(u, v) + iR_2(u, v), \qquad a = A_1(u, v) + iA_2(u, v),$$

that the electromagnetic field has generally a fixed curve of singularities. In the special case when

$$p = 1 - \tau^2, \quad q = 2\tau, \quad r = i(1 + \tau^2), \quad A_1 = h, \quad A_2 = k,$$

the fixed curve is the circle $x^2 + z^2 = \frac{1}{4}k^2, \quad y = \frac{1}{2}h$.

§ 48. **Cylindrical wave-functions with moving singularities.**

If we define τ in terms of x, y by the equation

$$[x - \xi(\tau)]^2 + [y - \eta(\tau)]^2 = c^2(t - \tau)^2,$$

and define $l(\tau)$, $m(\tau)$, $p(\tau)$, so that

$$l\xi'(\tau) + m\eta'(\tau) = c^2 p,$$
$$l^2 + m^2 = c^2 p^2,$$

the function $\quad \nu = l(x - \xi) + m(y - \eta) - c^2 p(t - \tau)$

is such that

$$\frac{1}{\sqrt{\nu}} f(\tau)$$

is a wave-function.

In particular, if $\xi = \eta = 0$, $l = 1$, $m = i$, $p = 0$ we obtain Poisson's wave-function

$$\frac{1}{\sqrt{x \pm iy}} f\left(t - \frac{\rho}{c}\right).$$

Another interesting result is that if $F(x, y, t)$ is a homogeneous function of degree $\frac{1}{2}$ which satisfies the wave-equation, the function

$$\Omega = F(x - \xi, y - \eta, t - \tau)$$

also satisfies the wave-equation.

If $\quad \sigma = \xi'(\tau)(x - \xi) + \eta'(\tau)(y - \eta) - c^2(t - \tau),$

the function

$$\Omega = \frac{1}{\sqrt{\sigma}} f(\tau)$$

satisfies the wave-equation only when $\xi'^2+\eta'^2=c^2$. When iz is written in place of ct these wave-functions can all be regarded as solutions of Laplace's equation.

EXAMPLES.

1. Let

$$\bar{x}=xf(\tau)-\int_0^\tau f'(s)\,\xi(s)\,ds, \qquad \bar{z}=zf(\tau)-\int_0^\tau f'(s)\,\zeta(s)\,ds,$$

$$\bar{y}=yf(\tau)-\int_0^\tau f'(s)\,\eta(s)\,ds, \qquad \bar{t}=tf(\tau)-\int_0^\tau f'(s)\,s\,ds;$$

where τ is defined in terms of x, y, z, t by means of equation (265) and the inequality $\tau \leqslant t$. Prove that if an electromagnetic field (E, H) is such that

$$\frac{1}{c}\frac{\partial\tau}{\partial t}\left(\frac{1}{c}\frac{\partial\tau}{\partial t}M_x \mp i\frac{\partial\tau}{\partial y}M_z \pm i\frac{\partial\tau}{\partial z}M_y\right)=\frac{\partial\tau}{\partial x}\left(\frac{\partial\tau}{\partial x}M_x+\frac{\partial\tau}{\partial y}M_y+\frac{\partial\tau}{\partial z}M_z\right),$$

....

....

with similar equations, where $M=H\pm iE$, then an electromagnetic field in the variables $\bar{x}, \bar{y}, \bar{z}, \bar{t}$ can be found such that we have identically

$$\begin{aligned} &M_x d(y, z)+M_y d(z, x)+M_z d(x, y)\mp icM_x d(x, t)\\ &\qquad\mp icM_y d(y, t)\mp icM_z d(z, t)\\ &\equiv \overline{M}_x d(\bar{y}, \bar{z})+\overline{M}_y d(\bar{z}, \bar{x})+\overline{M}_z d(\bar{x}, \bar{y})\\ &\qquad\mp ic\overline{M}_x d(\bar{x}, \bar{t})\mp ic\overline{M}_y d(\bar{y}, \bar{t})\mp ic\overline{M}_z d(\bar{z}, \bar{t}), \end{aligned}$$

where $d(y, z)$ denotes $dy\,\delta z-dz\,\delta y$ and dx, δx, etc. are two independent sets of increments of the variables.

2. Prove that if $f'(\tau)$ is always positive the variable $\bar{t}$ increases with t. Show also that if $(\bar{\xi}, \bar{\eta}, \bar{\zeta}, \bar{\tau})$ correspond to (ξ, η, ζ, τ), the point $(\bar{\xi}, \bar{\eta}, \bar{\zeta})$ moves along a curve $\bar{\Gamma}$ with a velocity less than that of light and that the velocities at two corresponding points of Γ and $\bar{\Gamma}$ are the same in magnitude and direction.

3. Prove that the conditions imposed upon the electromagnetic field in Example 1 are all satisfied in the case of the field specified by the potentials (266). Hence show that the transformation transforms the field of an electron moving along the curve Γ into the field of an electron moving along the curve $\bar{\Gamma}$.

4. Prove that the conditions of Example 1 are also satisfied for any electromagnetic field of type (282) where α and β are defined as in § 44. Hence show that a field of this type is transformed into another field of the same type associated with the curve $\bar{\Gamma}$.

5. Prove that the conditions of Example 1 are equivalent to only two conditions which imply that the field (E, H) is conjugate to any electromagnetic field of type (282).

6. Prove that electromagnetic fields of types (302) and (303) can be transformed into fields of the same general types with the aid of the transformation

$$\bar{x}=x+\int^{\tau} l(s)f(s)\,ds, \qquad \bar{z}=z+\int^{\tau} n(s)f(s)\,ds,$$

$$\bar{y}=y+\int^{\tau} m(s)f(s)\,ds, \qquad \bar{t}=t+\int^{\tau} p(s)f(s)\,ds.$$

7. If τ be defined in terms of x, y, z, t by the equations

$$[x-\xi(a,\beta,\tau)]^2+[y-\eta(a,\beta,\tau)]^2+[z-\zeta(a,\beta,\tau)]^2=c^2[t-\theta(a,\beta,\tau)]^2,$$

$$\frac{\partial\xi}{\partial a}(x-\xi)+\frac{\partial\eta}{\partial a}(y-\eta)+\frac{\partial\zeta}{\partial a}(z-\zeta)=c^2\frac{\partial\theta}{\partial a}(t-\tau),$$

$$\frac{\partial\xi}{\partial\beta}(x-\xi)+\frac{\partial\eta}{\partial\beta}(y-\eta)+\frac{\partial\zeta}{\partial\beta}(z-\zeta)=c^2\frac{\partial\theta}{\partial\beta}(t-\tau),$$

it satisfies the partial differential equation

$$\left(\frac{\partial\tau}{\partial x}\right)^2+\left(\frac{\partial\tau}{\partial y}\right)^2+\left(\frac{\partial\tau}{\partial z}\right)^2=\frac{1}{c^2}\left(\frac{\partial\tau}{\partial t}\right)^2.$$

8. If $\quad x=\rho\cos\phi, \quad y=\rho\sin\phi, \quad r^2=x^2+y^2+z^2,$

and

$$\sinh\psi=\frac{r^2-a^2}{2a\rho},$$

the function $\quad [x^2+y^2+(z\pm a)^2]^{-\frac{1}{2}}f[\phi+i\psi]$

satisfies Laplace's equation.

9. Prove that if in the last example we write

$$f[\phi+i\psi]=\frac{e^{iu}}{\phi+i\psi-ke^{iu}}$$

and integrate with regard to u between 0 and 2π we can obtain a potential function which is zero outside the tube $|\phi+i\psi|=k$.

10. Particles are projected in certain directions from the different positions of a moving electron and travel along straight lines with the velocity of light. Prove that if the law, according to which the direction of projection varies with the velocity of the electron, be suitably chosen the particles will at each instant form a line of electric force in the electromagnetic field due to the moving electron.

CHAPTER IX

MISCELLANEOUS THEORIES

§ 49. Kirchhoff's formula and its extensions.

An important solution of the wave-equation is embodied in Kirchhoff's formula* which is usually interpreted as the mathematical expression of the principle of Huygens. This formula has been extended by Love† and Macdonald‡ so as to give a representation of an electromagnetic field outside a surface in terms of the electric and magnetic forces tangential to the surface. In Macdonald's formula it is the time derivatives of E and H that are so expressed. Tonolo§ has given a formula in which E and H are expressed in terms of their surface values. The formulae are given in examples 3—5 at the end of this chapter.

When the surface is a sphere Kirchhoff's formula reduces to the formula of Poisson‖ (Ex. 5) which enables us to find a wave-function which satisfies the conditions

$$u = f(x, y, z), \qquad \frac{\partial u}{\partial t} = g(x, y, z).$$

Poisson's formula may be used to derive the theorem¶ that the mean value of a wave-function u over a sphere of radius $c\tau$ at time t is equal to the mean value of u at the centre of the

* *Berlin. Ber.* (1882), p. 641; *Wied. Ann.* Bd. 18 (1883); *Ges. Abh.* t. 2, p. 22. Simple proofs of the formula have been given by Beltrami, *Rend. Acc. Linc. Rom.* (5), t. 4 (1895); Larmor, *Proc. London Math. Soc.* Ser. 2, Vol. 1, p. 1; Love, *Ibid.* p. 37 (1903); Lamb, *Hydrodynamics*, 2nd edition (1906), p. 477; H. A. Lorentz, *The Theory of Electrons*, p. 233; E. Laura, *Il Nuovo Cimento* (1913).

† *Phil. Trans.* A, Vol. 197 (1901).

‡ *Electric Waves*, p. 16; *Proc. London Math. Soc.* Ser. 2, Vol. 10 (1911), p. 91; *Phil. Trans.* A, Vol. 212 (1912), p. 295. This theorem gives an analytical specification of a generalised Green's equivalent layer. See p. 29.

§ *Annali di Matematica*, Ser. 3, t. 17 (1910).

‖ The details of the calculation are given by Love, *loc. cit.* A simple proof of Poisson's formula is given by Lamb, *loc. cit.* p. 471.

¶ Cf. Rayleigh's *Sound*, appendix, and H. Bateman, *Amer. Journ.* (1912), where some other theorems of a similar kind are given.

sphere during the interval $t-\tau$ to $t+\tau$. The function u is subject to the conditions in Kirchhoff's theorem.

When the function u is independent of z, Poisson's formula reduces to Parseval's formula for a cylindrical wave-function. Volterra* has extended Parseval's formula so as to obtain a two-dimensional analogue of Kirchhoff's formula. His formula indicates that the propagation of cylindrical waves is essentially different in character from that of spherical waves. In the three-dimensional case the value of a wave-function $u(x, y, z, t)$ at a point (x, y, z) at time t is completely determined by the values of u and $\frac{\partial u}{\partial t}$ over a concentric sphere of radius $c\tau$ at time $t-\tau$. In the two-dimensional case, on the other hand, the value of $u(x, y, t)$ at a point (x, y) at time t is not determined by the values of u and $\frac{\partial u}{\partial t}$ over a concentric circle at time $t-\tau$. To find $u(x, y, t)$ we must know the values of u and $\frac{\partial u}{\partial t}$ over a series of such circles in which the radius $c\tau$ varies from 0 to some other value $c\tau_1$. The essential difference between the two cases may be attributed to the fact that in the three-dimensional case the wave-function for a source is of type $\frac{1}{r}f\left(t-\frac{r}{c}\right)$, while in the two-dimensional case it is of type $\int_0^\infty f\left(t-\frac{r}{c}\cosh u\right)$ and a wave does not leave the region undisturbed after it has passed, but has a tail or residue†.

When u is a periodic function of t, Kirchhoff's formula may be replaced by the simple formula of Helmholtz‡. In this case there is an analogous formula for cylindrical wave-functions, the function $K_0(i\rho k)$ taking the place of e^{ikr}/r.

§ 50. Green's Functions.

The solution of a problem in which a periodic wave-function is to be determined from a knowledge of its behaviour at

* *Acta Math.* t. 18; *Lectures at Clark University* (1912), p. 38.

† See Lamb's *Hydrodynamics*, p. 474.

‡ See also J. Hadamard, *Bull. de la Société math. de France*, t. 28 (1900), p. 69; J. Larmor, *Proc. London Math. Soc.* (2), Vol. 1 (1903), p. 13.

certain boundaries can be made to depend on that of an auxiliary problem, viz. the determination of the Green's function*.

Let $G(x, y, z; x_1, y_1, z_1)$ be a solution of $\Delta u + k^2 u = 0$ with the following properties: It is to be finite and continuous, as also its first and second derivatives, in a region bounded by a surface S, except in the neighbourhood of the point (x_1, y_1, z_1), where it is to be infinite like $\cos kr/4\pi r$, when $r \to 0$. At the surface S, G satisfies some boundary condition such as (1) $u = 0$ or (2) $\frac{\partial u}{\partial n} = 0$.

Adopting the notation of Plemelj † and Kneser‡ we shall denote the values of a function $\phi(\xi, \eta, \zeta)$ at the points (x, y, z), (x_1, y_1, z_1) respectively by $\phi(0)$ and $\phi(1)$. The Green's function is then denoted by the symbol $G(0, 1)$. The importance of the Green's function depends chiefly on the following theorem.

Let ϕ be a solution of

$$\Delta\phi + k^2\phi + f(x, y, z) = 0,$$

which is finite and continuous, together with its first and second derivatives, through the interior of the region and satisfies the same boundary condition as $G(0, 1)$, then

$$\phi(1) = \iiint f(0)\, G(0, 1)\, dx\, dy\, dz.$$

This theorem is proved by applying Green's theorem to the region between a small sphere Σ, whose centre is at (x_1, y_1, z_1), and the surface S. For since

$$\iiint (\phi\Delta G - G\Delta\phi)\, dx\, dy\, dz = \iint \left(\phi \frac{\partial G}{\partial n} - G \frac{\partial \phi}{\partial n}\right) dS - \iint \left(\phi \frac{\partial G}{\partial n} - G \frac{\partial \phi}{\partial n}\right) d\Sigma,$$

we obtain the required relation by making $\Sigma \to 0$ and using the boundary conditions.

* This function was first used by Green in the solution of a problem of electrostatics, *Essay on the application of mathematical analysis to the theories of electricity and magnetism*, Nottingham (1828); *Math. Papers*, p. 31.

† *Monatshefte für Math. u. Phys.* (1904) and (1907).

‡ *Die Integralgleichungen und ihre Anwendung in der mathematischen Physik*, § 31, Brunswick (1911).

If $g(2, 0)$ is the Green's function for the same boundary condition as $G(2, 0)$ but for $k = \sigma$, we must also surround the point (2) by a small sphere when we apply Green's theorem with $\phi(0) = g(2, 0)$. We then obtain the equation

$$g(2, 1) = G(2, 1) - (k^2 - \sigma^2) \iiint g(2, 0)\, G(0, 1)\, dx\, dy\, dz.$$

This important relation indicates that $g(2, 1)$ is the solving function of an integral equation of which $G(0, 1)$ is kernel and *vice versâ*. The theory of integral equations tells us that when $G(0, 1)$ is given there may be certain singular values of σ^2 for which $g(2, 1)$ is not finite. These are the values of σ^2 for which the homogeneous integral equation

$$\phi(1) = (\sigma^2 - k^2) \iiint \phi(0)\, G(0, 1)\, dx\, dy\, dz$$

possesses a continuous solution $\phi(0)$ which is different from zero. Formula (2) indicates that for such values of σ^2 the differential equation $\Delta\phi + k^2\phi = 0$ possesses a solution satisfying the boundary condition and the other conditions imposed on ϕ. The solutions of this type are of great importance in the theory of sound and have been discussed by many writers*.

If we put $f(0) = (\sigma^2 - k^2)\, g(0, 2)$ and proceed as before, Green's theorem gives

$$g(1, 2) = G(2, 1) - (k^2 - \sigma^2) \iiint g(0, 2)\, G(0, 1)\, dx\, dy\, dz.$$

Putting $\sigma = k$ and comparing this with the previous equation we get

$$g(1, 2) = g(2, 1).$$

Hence the Green's function is a symmetric function of the coordinates of the points 1, 2. When the boundary condition is $\frac{\partial u}{\partial n} = 0$ this result is equivalent to Helmholtz's theorem†.

Since G is a real symmetric function when $k = 0$ it follows from the general theory of integral equations that there is

* See especially Lord Rayleigh, *Theory of Sound*, Vol. 2; Pockels, *Die partielle Differentialgleichung* $\Delta u + k^2 u = 0$; A. Sommerfeld, *Encyklopädie der Math. Wiss.* Band II. 1, Heft 4, p. 540.

† Cf. Rayleigh's *Sound*, Vol. 2, p. 131.

at least one real singular value of σ^2; that all the singular values are positive may be deduced at once from the equation

$$\sigma^2 \iiint \phi^2\, dxdydz = -\iiint \phi\Delta\phi\, dxdydz$$
$$= \iiint \left[\left(\frac{\partial\phi}{\partial x}\right)^2 + \left(\frac{\partial\phi}{\partial y}\right)^2 + \left(\frac{\partial\phi}{\partial z}\right)^2\right] dxdydz.$$

The Green's function is usually obtained in practice by finding a suitable expansion in terms of elementary solutions of the equation $\Delta u + k^2 u = 0$. This method is explained in Heine's *Kugelfunktionen* and many examples of Green's functions are given for the case $k = 0$. The general case has been discussed at length by A. Sommerfeld* who also obtains a number of definite integrals which represent Green's functions. These expressions lead to interesting generalisations of Fourier's theorem.

The problem of electrical oscillations in a cavity has been discussed by Weyl†. With the aid of a generalisation of the Green's function, viz. a Green's tensor, he obtains a number of inequalities satisfied by the periods of vibration.

The Green's function for the equation $\Delta u + k^2 u = 0$ can theoretically be found when the corresponding Green's function for the equation $\Delta u = 0$ is known. Considerable progress has been made in the theory since the appearance of Heine's work and so a few references to recent literature will be useful‡. A

* *Phys. Zeitschr.* Bd. 11 (1910), p. 1087; *Jahresbericht der deutsch. math. Verein*, Bd. 21 (1913).

† *Math. Ann.* Bd. 71, p. 441; *Crelle*, Bd. 141 (1912).

‡ For the determination of special Green's functions see E. W. Hobson, *Cambr. Phil. Trans.* Vol. 18 (1899), p. 277; H. M. Macdonald, *Ibid.* p. 292, *Proc. London Math. Soc.* Vol. 26 (1895), p. 161; A. G. Greenhill, *Proc. Cambr. Phil. Soc.* Vol. 3 (1880); J. Dougall, *Proc. Edinburgh Math. Soc.* (1900); H. S. Carslaw, *Ibid.* (1912), *Proc. London Math. Soc.* (2), Vol. 8, p. 365; C. W. Oseen, *Arkiv för mat.* Bd. 2; C. Neumann, *Leipziger Berichte*, Bd. 58 (1906), Bd. 62 (1910); W. Burnside, *Proc. London Math. Soc.* Vol. 25 (1894), p. 94. For the general theory H. Poincaré, *Rend. Palermo*, t. 8 (1894), p. 57; S. Zaremba, *Ibid.* t. 19 (1905); E. R. Neumann, *Studien über die Methode von C. Neumann und G. Robin zur Lösung der beiden dwertaufgaben der Potentialtheorie*, Leipzig (1905); D. Hilbert, *Gött. Nachr.* 1904); M. Mason, *Newhaven Math. Colloquium* (1910); E. Picard, *Ann. de l'École Normale* (1906), p. 509.

good account of the developments up to 1900 is given in Sommerfeld's article in the *Encyklopädie der Mathematischen Wissenschaften.*

§ 51. The transformation of the electromagnetic equations.

The transformations which can be used to transform any solution of the wave-equation into another solution or any electromagnetic field into another belong to a group which is characterised by a relation of the form*

$$dx'^2 + dy'^2 + dz'^2 - c^2dt'^2 = \lambda^2 (dx^2 + dy^2 + dz^2 - c^2dt^2).$$

The linear transformations belonging to this group are of great importance in the modern theory of relativity†; two of the non-linear transformations have been mentioned in § 13.

In addition to these transformations there are other transformations, involving arbitrary functions in their specification, which can be applied to certain types of wave-functions, and to certain types of electromagnetic fields. There are often two families of wave-functions to which a given transformation can be applied, when the transformation is of a suitable character; each of these families may be defined by a linear relation which exists between the wave-function and its derivatives, sometimes between the derivatives alone. Some idea of the theory may be derived from the examples. It also happens that there is often a family of electromagnetic fields to which a given transformation can be applied and this family is defined by means of two linear relations between E and H, which can be interpreted to mean that the field is conjugate to some definite electromagnetic field or family of electromagnetic fields determined by the transformation. In some cases these last fields are self-conjugate and the transformation is applicable to them also.

* H. Bateman, *Proc. London Math. Soc.* Ser. 2, Vol. 7 (1909), Vol. 8 (1910); E. Cunningham, *Ibid.* Vol. 8 (1910).

† For this see A. Einstein, *Ann. d. Phys.* Bd. 17 (1905); Laue, *Das Relativitätsprinzip*, Brunswick (1911); E. Cunningham, *British Association Reports* (1911); H. Minkowski, *Gött. Nachr.* (1908); E. B. Wilson and G. N. Lewis, *Proc. Amer. Acad. of Arts and Sciences*, Vol. 48 (1912), p. 389; J. Ishiwara, "Bericht über die Relativitätstheorie," *Jahrbuch der Radioaktivität und Elektronik*, Bd. 9 (1912), pp. 560—648; L. Silberstein, *The Theory of Relativity*, Macmillan and Co. (1914).

The fact that the condition of conjugacy between two electromagnetic fields often implies the existence of one or more transformations depending on arbitrary functions, may be regarded as of some philosophical interest.

MISCELLANEOUS EXAMPLES.

1. Show that the most general periodic solution ξe^{iat} (valid for all space outside a given closed surface) of the wave-equation is

$$\xi = \int \frac{e^{iar}}{r} \left\{ P + Q \cos\psi \left(ia - \frac{1}{r} \right) \right\} dS,$$

where P and Q are arbitrary functions, r is the distance from the element of surface dS to the point where ξ is estimated, and ψ is the angle between r and the outward drawn normal. Show further that the necessary and sufficient condition that the value of ξ, given by the same analytical expression, should vanish for points inside the surface, is that $P = \frac{\partial Q}{\partial n}$.

(Cambr. Math. Tripos, Part II, 1904.)

2. Let Ω be a function which satisfies the wave-equation and is such that its differential coefficients of the first order are continuous functions of x, y, z, t within a region bounded by a closed surface S. If either Ω or $\frac{\partial \Omega}{\partial n}$ be given for points on the surface S there is only one function Ω which reduces to a given function $f(x, y, z)$ for $t = t_0$.

(A. E. H. Love, *Proc. London Math. Soc.* Ser. 2, Vol. 1, p. 42; J. Hadamard, *Bull. de la Soc. Math. de France*, t. 28 (1900).)

3. If throughout a specified region of space and a specified interval of time u and its differential coefficients of the first order are continuous functions of x, y, z and of t, if also the differential coefficients of the second order such as $\frac{\partial^2 u}{\partial t^2}, \frac{\partial^2 u}{\partial x^2}$ are finite and integrable, then a solution of the equation

$$\Omega u + \sigma(x, y, z, t) = 0$$

which is valid for this region is given by the formula

$$u(x_0, y_0, z_0, t_0) = \frac{1}{4\pi} \iint \left\{ [u] \frac{\partial}{\partial n}\left(\frac{1}{r}\right) - \frac{1}{r}\left[\frac{\partial u}{\partial n}\right] - \frac{1}{r}\frac{\partial r}{\partial n}\left[\frac{\partial u}{\partial t}\right] \right\} dS$$
$$+ \frac{1}{4\pi} \iiint \frac{[\sigma]}{r}\, dx\, dy\, dz,$$

where $r^2 = (x - x_0)^2 + (y - y_0)^2 + (z - z_0)^2$, n denotes the normal to dS drawn into the specified region and the integration is taken throughout this region and over its boundary. The function σ is supposed to be finite and integrable and a quantity within square brackets is calculated at time

$$t=t_0-\frac{r}{c}. \qquad \text{(G. Kirchhoff.)}$$

4. If an electromagnetic field is such that the specified region does not contain any charges or convection currents and $M=H+iE$, the value M^0 of the vector M at (x_0, y_0, z_0, t_0) is given by the formula

$$4\pi M_x^0=\iint\{[M_r]\cos\widehat{nx}-[M_n]\cos\widehat{rx}-[M_x]\cos\widehat{rn}\}\frac{dS}{r^2}$$
$$+\frac{1}{c}\frac{\partial}{\partial t_0}\iint\{[M_r]\cos\widehat{nx}-[M_n]\cos\widehat{rx}-[M_x]\cos\widehat{rn}$$
$$+i\,[M_z]\cos\widehat{ny}-i\,[M_y]\cos\widehat{nz}\}\frac{dS}{r}. \qquad \text{(A. Tonolo.)}$$

5. Prove that in the same circumstances

$$4\pi\frac{\partial M_x^0}{\partial t_0}=\frac{1}{c}\frac{\partial}{\partial t_0}\iint\left[\frac{\partial}{\partial y_0}\left(\frac{\gamma}{r}\right)-\frac{\partial}{\partial z_0}\left(\frac{\beta}{r}\right)\right]dS$$
$$+i\iint\left[\frac{\partial^2}{\partial x_0^2}\left(\frac{\alpha}{r}\right)+\frac{\partial^2}{\partial x_0\partial y_0}\left(\frac{\beta}{r}\right)+\frac{\partial^2}{\partial x_0\partial z_0}\left(\frac{\gamma}{r}\right)-\frac{1}{c^2}\frac{\partial^2}{\partial t_0^2}\left(\frac{\alpha}{r}\right)\right]dS,$$

where $\alpha=\mu[M_z]-\nu[M_y]$, etc. and (λ, μ, ν) are the direction cosines of the normal drawn into the region bounded by S.

(H. M. Macdonald.)

6. If u is a wave-function independent of z and periodic in t like e^{ikt}

$$u(x_0, y_0)=\frac{1}{2\pi}\int\left\{K_0(ikr)\frac{\partial u}{\partial n}-u\frac{\partial}{\partial n}K_0(ikr)\right\}ds.$$

7. A wave-function which satisfies the conditions

$$u=f(x, y, z) \qquad \frac{\partial u}{\partial t}=g(x, y, z)$$

is given by the formula

$$u=\frac{\partial}{\partial t}(t\bar{f})+t\bar{g},$$

where $\bar{f}$, $\bar{g}$ denote the mean values of f, g respectively over the surface of a sphere of radius ct having the point x, y, z as centre.

(S. D. Poisson.)

8. If u satisfies $\frac{\partial^2 u}{\partial x^2}+\frac{\partial^2 u}{\partial y^2}=\frac{\partial^2 u}{\partial t^2}$ and has finite second derivatives within a suitable domain

$$u(x_1, y_1, t_1)=\frac{1}{2\pi}\frac{\partial}{\partial t_1}\int_\sigma\frac{d\sigma}{\sqrt{(t_1-t)^2-\rho^2}}u(x, y, t)$$
$$+\frac{1}{2\pi}\int_\sigma\frac{d\sigma}{\sqrt{(t_1-t)^2-\rho^2}}\frac{\partial}{\partial t}u(x, y, t),$$

where $\rho^2=(x-x_1)^2+(y-y_1)^2$,

and σ denotes the area within the circle cut out on the plane $T=t$ by the cylinder

$$(X-x_1)^2+(Y-y_1)^2=(T-t_1)^2.$$

(Parseval and Volterra.)

9. Prove that if a transformation of variables from (x, y, z, t) to $(\bar{x}, \bar{y}, \bar{z}, \bar{t})$ is such that

$$d\bar{x}^2+d\bar{y}^2+d\bar{z}^2-c^2d\bar{t}^2=\lambda^2(dx^2+dy^2+dz^2-c^2dt^2)+(l\,dx+m\,dy+n\,dz-cp\,dt)$$
$$(l_0dx+m_0dy+n_0dz-cp_0dt),$$

where $l^2+m^2+n^2=p^2$, $l_0{}^2+m_0{}^2+n_0{}^2=p_0{}^2$; it can be used to transform an electromagnetic field (E, H) into another electromagnetic field $(\bar{E}, \bar{H})$ with an identical relation of the same type as that used in Ex. 1, Ch. VIII, if the two conditions embodied in the relation

$$M_x(mn_0-m_0n)+M_y(nl_0-n_0l)+M_z(lm_0-l_0m)\mp iM_x(lp_0-l_0p)$$
$$\mp iM_y(mp_0-m_0p)\mp iM_z(np_0-n_0p)=0$$

are satisfied.

Prove that the conditions can also be thrown into the form

$$(ll_0+mm_0+nn_0+pp_0)\,M_x\mp i(np_0+n_0p)\,M_y\pm i(mp_0+m_0p)\,M_z$$
$$=l_0(lM_x+mM_y+nM_z)+l\,(l_0M_x+m_0M_y+n_0M_z)$$

and similar equations.

10. In the last example if $l_0=l$, $m_0=m$, $n_0=n$, $p_0=-p$, the conditions are satisfied if Poynting's vector is in the direction (l, m, n). Show, in particular, that the transformation

$$\bar{x}\pm i\bar{y}=F\left(\frac{x\pm iy}{z+r}\right),\quad \bar{z}=f(r+ct)+g(r-ct),\quad c\bar{t}=f(r+ct)-g(r-ct)$$

can be applied to an electromagnetic field in which Poynting's vector is along the radius from the origin and that in the resulting electromagnetic field Poynting's vector is parallel to the axis of z. Apply the transformation to the electromagnetic field derived from the functions α, β given by equations (13), § 5.

11. If a transformation of coordinates is such that

$$d\bar{x}^2+d\bar{y}^2+d\bar{z}^2-c^2d\bar{t}^2=\lambda^2(dx^2+dy^2+dz^2-c^2dt^2),$$

where λ is a function of x, y, z, t; there is also a relation of type

$$(\bar{x}-\bar{x}_0)^2+(\bar{y}-\bar{y}_0)^2+(\bar{z}-\bar{z}_0)^2-c^2(\bar{t}-\bar{t}_0)^2$$
$$=\lambda\lambda_0[(x-x_0)^2+(y-y_0)^2+(z-z_0)^2-c^2(t-t_0)^2].$$

(J. Liouville and S. Lie.)

12. Prove that the differential equation

$$\left(\frac{\partial\tau}{\partial x}\right)^2+\left(\frac{\partial\tau}{\partial y}\right)^2+\left(\frac{\partial\tau}{\partial z}\right)^2=\frac{1}{c^2}\left(\frac{\partial\tau}{\partial t}\right)^2$$

is covariant for a transformation of the type considered in Ex. 11.

(S. Lie.)

13. Prove that if ξ, η, ζ, τ are functions of x, y, z, t such that

$$(\zeta M_y - \eta M_z \pm ic\tau M_x)\, dx + (\xi M_z - \zeta M_x \pm ic\tau M_y)\, dy$$
$$+ (\eta M_x - \xi M_y \pm ic\tau M_z)\, dz \mp ic\,(\xi M_x + \eta M_y + \zeta M_z)\, dt$$

is an exact differential and ϵ is a quantity whose square may be neglected, the value of M at the point $x' = x + \epsilon\xi$, $y' = y + \epsilon\eta$, $z' = z + \epsilon\zeta$, $t' = t + \epsilon\tau$ may be calculated by assuming that the integral form

$$M_x(dy\,\delta z - dz\,\delta y) + M_y(dz\,\delta x - dx\,\delta z) + M_z(dx\,\delta y - dy\,\delta x)$$
$$\mp icM_x(dx\,\delta t - dt\,\delta x) \mp icM_y(dy\,\delta t - dt\,\delta y) \mp icM_z(dz\,\delta t - dt\,\delta z)$$

is an invariant for the infinitesimal transformation, it being supposed that the function M satisfies equations (10) of § 5.

14. Let a transformation from the coordinates (x', y', z', t') to (x, y, z, t) be such that

$$dx'^2 + dy'^2 + dz'^2 - c^2 dt'^2 = \sqrt{\Lambda}\,(dx^2 + dy^2 + dz^2 - c^2 dt^2)$$
$$+ \frac{\Theta'}{\sqrt{\Lambda}}(l\,dx + m\,dy + n\,dz - c^2 p\,dt)^2$$
$$- \frac{2\Phi}{\sqrt{\Lambda}}(l\,dx + m\,dy + n\,dz - c^2 p\,dt)(l'dx + m'dy + n'dz - c^2 p'dt)$$
$$+ \frac{\Theta}{\sqrt{\Lambda}}(l'dx + m'dy + n'dz - c^2 p'dt)^2,$$

where $\Theta = l^2 + m^2 + n^2 - c^2p^2$, $\quad \Theta' = l'^2 + m'^2 + n'^2 - c^2 p'^2$,

$\Phi = \sigma + ll' + mm' + nn' - c^2 pp'$, $\quad \Lambda = \Phi^2 - \Theta\Theta'$;

and σ, l, m, n, p, l', m', n', p' are functions of x, y, z, t; then if θ satisfies the equations

$$l\frac{\partial\theta}{\partial x} + m\frac{\partial\theta}{\partial y} + n\frac{\partial\theta}{\partial z} + p\frac{\partial\theta}{\partial t} + k\theta = 0,$$

$$\frac{\partial}{\partial x}\left[\sigma\frac{\partial\theta}{\partial x} - (\alpha + 2l'k)\,\theta\right] + \frac{\partial}{\partial y}\left[\sigma\frac{\partial\theta}{\partial y} - (\beta + 2m'k)\,\theta\right] + \frac{\partial}{\partial z}\left[\sigma\frac{\partial\theta}{\partial z} - (\gamma + 2n'k)\,\theta\right]$$
$$= \frac{1}{c^2}\frac{\partial}{\partial t}\left[\sigma\frac{\partial\theta}{\partial t} - c^2(\epsilon - 2p'k)\,\theta\right],$$

it also satisfies $\quad \dfrac{\partial^2\theta}{\partial x'^2} + \dfrac{\partial^2\theta}{\partial y'^2} + \dfrac{\partial^2\theta}{\partial z'^2} - \dfrac{1}{c^2}\dfrac{\partial^2\theta}{\partial t'^2}.$

In the preceding equation we have

$$\alpha = \frac{\partial R}{\partial y} - \frac{\partial Q}{\partial z} + \frac{\partial L}{\partial t}, \qquad \beta = \frac{\partial P}{\partial z} - \frac{\partial R}{\partial x} + \frac{\partial M}{\partial t},$$
$$\gamma = \frac{\partial Q}{\partial x} - \frac{\partial P}{\partial y} + \frac{\partial N}{\partial t}, \qquad \epsilon = \frac{\partial L}{\partial x} + \frac{\partial M}{\partial y} + \frac{\partial N}{\partial z},$$
$$P = mn' - m'n, \qquad Q = nl' - n'l, \qquad R = lm' - l'm,$$
$$L = lp' - l'p, \qquad M = mp' - m'p, \qquad N = np' - n'p.$$

Show that in certain cases a function of type $\lambda\theta$ is a solution of the wave-equation in consequence of the two equations imposed on θ. Discuss the case of the transformation of Ex. 1, p. 139: also the case of a transformation which leaves the functions X, Y, Z of Ex. 7, p. 80 unaltered in form.

15. Prove that if a function V satisfies the equation

$$\frac{\partial^2 V}{\partial x^2}+\frac{\partial^2 V}{\partial y^2}=\frac{1}{c^2}\frac{\partial^2 V}{\partial t^2},$$

vanishes at infinity and has continuous derivatives except at points of the curve σ where its normal derivative is discontinuous in such a way that

$$\frac{\partial V}{\partial n}+\frac{\partial V}{\partial n'}=-f(Q, t),$$

the symbol Q being used to denote the coordinates of a point Q, then

$$V(P, t)=\frac{1}{2\pi}\int_\sigma d\sigma\int_\rho^{ct} f\left(Q, t-\frac{s}{c}\right)\frac{ds}{\sqrt{s^2-\rho^2}},$$

where $\rho^2=x^2+y^2$.

(Levi-Cività, *Nuovo Cimento*, 1897.)

16. Prove that

$$\frac{1}{\sqrt{(x-\xi)^2+(y-\eta)^2+(z-\zeta)^2}}$$
$$=\frac{1}{2\pi^2}\int_{-\infty}^{\infty}\int_{-\infty}^{\infty}\int_{-\infty}^{\infty}\frac{e^{i\lambda(x-\xi)+i\mu(y-\eta)+i\nu(z-\zeta)}}{\lambda^2+\mu^2+\nu^2}\,d\lambda\,d\mu\,d\nu,$$

and deduce that the integral $V=\iiint\frac{1}{r}\rho(\xi,\eta,\zeta)\,d\xi\,d\eta\,d\zeta$ satisfies Poisson's equation $\Delta V+4\pi\rho(x, y, z)=0$.

(J. Weingarten.)

17. Prove that the equation $\Delta\phi=\frac{\partial^2\phi}{\partial t^2}+2\frac{\partial\phi}{\partial t}$ is satisfied by

$$\phi=e^{-t}\left[\{f(t-t_1-r)+f(t-t_1+r)\}\frac{1}{r}+\int_{-\infty}^{t-t_1-r} f(\tau)\,I_0'(\theta)\frac{d\tau}{\theta}\right.$$
$$\left.+\int_{t-t_1+r}^{\infty} f(\tau)\,I_0'(\theta)\frac{d\tau}{\theta}\right],$$

where $\theta^2=(t-t_1-\tau)^2-r^2,\quad r^2=(x-x_0)^2+(y-y_0)^2+(z-z_0)^2$,

and $I_0(\theta)$ is the Bessel's function $J_0(i\theta)$. x_0, y_0, z_0 and t_1 are arbitrary constants.

(M. Brillouin, *Comptes Rendus*, 1903.

18. Prove that a solution of the differential equation

$$\frac{\partial^2 U}{\partial t^2}-\frac{\partial^2 U}{\partial x^2}=U$$

is given by

$$U=\int_{-\infty}^{\infty} e^{i\lambda x}\left[F(\lambda)\cos(t\sqrt{\lambda^2-1})+G(\lambda)\frac{\sin(t\sqrt{\lambda^2-1})}{\sqrt{\lambda^2-1}}\right]d\lambda;$$

hence obtain a solution of the equation $\frac{\partial^2 V}{\partial t^2}+2\frac{\partial V}{\partial t}=\frac{\partial^2 V}{\partial x^2}$ by putting $V=U(x,t)e^{-t}$. If $U=f(x), \frac{\partial U}{\partial t}=g(x)$ when $t=0$, the functions F, G may be determined by the equations

$$f(x)=\int_{-\infty}^{\infty} e^{i\lambda x} F(\lambda)\, d\lambda, \qquad g(x)=\int_{-\infty}^{\infty} e^{i\lambda x} G(\lambda)\, d\lambda,$$

with the aid of Fourier's theorem.

(H. Poincaré, *Comptes Rendus*, 1893–4.)

19. Prove that with the conditions of the last example

$$U(x, t)=\tfrac{1}{2}[f(x-t)+f(x+t)]+\tfrac{1}{2}\int_{x-t}^{x+t}\left[f(\xi)\frac{\partial G}{\partial \tau}-g(\xi)G\right]_{\tau=0} d\xi,$$

where $\qquad G(x, \xi;\ t, \tau)=J_0\sqrt{(t-\tau)^2-(x-\xi)^2}.$

(Laplace (1779); E. Picard, *Bull. soc. math.* t. 22 (1894), p. 2.)

20. Prove that a solution $\Phi=e^{-t}V$ of the equation $\Delta\Phi=\frac{\partial^2\Phi}{\partial t^2}+2\frac{\partial\Phi}{\partial t}$ is given by the following extension of Kirchhoff's formula:

$$\begin{aligned}4\pi V(t_1, x_0, y_0, z_0)=&\iiint\left[G\frac{\partial V}{\partial t}-V\frac{\partial G}{\partial t}\right]_{t=0} dx\,dy\,dz\\ &+\iint_{\text{sphere } r=t_1}\left[\tfrac{1}{2}V+\frac{1}{r}\left(\frac{\partial V}{\partial t}+\frac{\partial V}{\partial r}\right)+\frac{1}{r^2}V\right]_{t=0} dS\\ &-\iint_{\Sigma}\left\{\left[VG\frac{\partial r}{\partial n}\right]_{t=0}+\left[\frac{1}{r}\left(\frac{\partial V}{\partial t}\frac{\partial r}{\partial n}+\frac{\partial V}{\partial n}\right)+\frac{V}{r^2}\frac{\partial r}{\partial n}\right]_{t=t_1-r}\right\}d\Sigma\\ &-\iint_{\Sigma} d\Sigma\int_0^{t_1-r}\left[G\left(\frac{\partial V}{\partial n}+\frac{\partial V}{\partial t}\frac{\partial r}{\partial n}\right)+V\frac{\partial r}{\partial n}\left(\frac{\partial G}{\partial t}-\frac{\partial G}{\partial r}\right)\right]dt;\end{aligned}$$

where $r^2=(x-x_0)^2+(y-y_0)^2+(z-z_0)^2$, $\theta^2=(t-t_1)^2-r^2$, $G=\frac{1}{\theta}\frac{d}{d\theta}I_0(\theta)$ and $I_0(\theta)$ is the Bessel's function $J_0(i\theta)$.

The first integral extends over the volume enclosed by both the sphere $r=t_1$ and the surface Σ; when this sphere cuts the surface the second integral extends over the part of the spherical surface inside Σ, the last two integrals extend over the part of Σ which lies inside the sphere. The normal n is supposed to be drawn into the region of integration. If Σ is a closed surface and (x_0, y_0, z_0) lies outside, the region of integration is the space outside Σ and inside the sphere.

(M. Brillouin, *Comptes Rendus*, 1903.)

21. Let α, β, ω be defined in terms of x, y, z by the equations

$$\gamma^2 x=\cos\alpha\cos(\beta+k\omega), \quad \gamma^2 y=\cos\alpha\sin(\beta+k\omega), \quad \gamma^2 z=\sin\alpha\cos\omega,$$

where $\gamma^2=1-\sin\alpha\sin\omega$ and k is a constant. A solution of Laplace's equation $\Delta u=0$ is then given by $u=\gamma F(a,\beta)$ provided F satisfies the partial differential equation

$$\sin 2a\frac{\partial^2 F}{\partial a^2}+4\frac{\sin^2 a+k^2\cos^2 a}{\sin 2a}\frac{\partial^2 F}{\partial \beta^2}+2\cos 2a\frac{\partial F}{\partial a}-\frac{3}{4}F\sin 2a=0.$$

(U. Amaldi, *Rend. Palermo* (1902), p. 1.)

22. Prove that the following transformations of coordinates lead to binary potentials, i.e. to solutions of Laplace's equation of type $u=F(\xi,\eta)$:

(1) $x=\xi,\quad y=\eta,\quad z=\zeta;$
(2) $x=\xi\cos\zeta,\quad y=\xi\sin\zeta,\quad z=\eta;$
(3) $x=\xi\cos\zeta,\quad y=\xi\sin\zeta,\quad z=\eta-m\zeta;$
(4) $x=\zeta\sin\xi\cos\eta,\quad y=\zeta\sin\xi\sin\eta,\quad z=\zeta\cos\xi;$
(5) $x=\xi\sin\eta e^{m\zeta}\cos\zeta,\quad y=\xi\sin\eta e^{m\zeta}\sin\zeta,\quad z=\xi\cos\eta e^{m\zeta};$

where m is an arbitrary constant. The differential equations satisfied by F in cases (3) and (5) are

$$\frac{\partial^2 F}{\partial \xi^2}+\left(1+\frac{m^2}{\xi^2}\right)\frac{\partial^2 F}{\partial \eta^2}+\frac{1}{\xi}\frac{\partial F}{\partial \xi}=0$$

and

$$(1+m^2\operatorname{cosec}^2\eta)\frac{\partial^2 F}{\partial \xi^2}+\frac{1}{\xi^2}\frac{\partial^2 F}{\partial \eta^2}+\frac{1}{\xi}(2+m^2\operatorname{cosec}^2\eta)\frac{\partial F}{\partial \xi}+\frac{1}{\xi^2}\cot\eta\frac{\partial F}{\partial \eta}=0$$

respectively. In the other cases the differential equations are already familiar.

(T. Levi-Cività, *Turin Memoirs*, (2) t. 49 (1900).)

23. If the differential equation $\dfrac{\partial^2 V}{\partial x^2}+\dfrac{\partial^2 V}{\partial y^2}=\dfrac{\partial^2 V}{\partial t^2}$ is satisfied by an expression of type $V=\gamma f(\theta)$ where f is an arbitrary function, θ must satisfy the differential equation

$$\left(\frac{\partial\theta}{\partial x}\right)^2+\left(\frac{\partial\theta}{\partial y}\right)^2=\left(\frac{\partial\theta}{\partial t}\right)^2.$$

Prove that if we write $\mu d\theta=\cos a\,dx+\sin a\,dy-dt$, $x=t\cos a+u$, $y=t\sin a+v$, there is a relation between a, u, v. Discuss the cases in which a is a function of θ and a constant respectively, and obtain the general value of γ in each case.

24. If in the last example $u=f(a,\theta)$, $v=g(a,\theta)$ and we write

$$\frac{\partial f}{\partial a}=\tau\sin a,\quad \frac{\partial g}{\partial a}=-\tau\cos a,$$

$$\frac{\partial f}{\partial \theta}=\sigma\cos\epsilon,\quad \frac{\partial g}{\partial \theta}=\sigma\sin\epsilon,$$

the new form of the equation $\frac{\partial^2 V}{\partial x^2}+\frac{\partial^2 V}{\partial y^2}=\frac{\partial^2 V}{\partial t^2}$ when a, θ, t are taken as independent variables is

$$\frac{\partial}{\partial a}\left[\frac{\sigma}{t-\tau}\cos(a-\epsilon)\frac{\partial V}{\partial a}+\sigma\sin(a-\epsilon)\frac{\partial V}{\partial x}\right]+\frac{\partial}{\partial \theta}\left[(t-\tau)\frac{\partial V}{\partial t}\right]$$

$$+\frac{\partial}{\partial t}\left[(t-\tau)\frac{\partial V}{\partial \theta}-\sigma(t-\tau)\cos(a-\epsilon)\frac{\partial V}{\partial t}\right]=0.$$

Prove that this equation can only possess a solution of type $V=\gamma F(\theta)$, with F arbitrary, if $\frac{\partial \tau}{\partial a}=0$, and in this case θ is defined by an equation of type

$$[x-\xi(\theta)]^2+[y-\eta(\theta)]^2=[t-\tau(\theta)]^2,$$

while the most general value of γ is

$$\frac{\{l(\theta)[x-\xi]+m(\theta)[y-\eta]+n(\theta)[t-\tau]\}^{\frac{1}{2}}}{\xi'(\theta)[x-\xi]+\eta'(\theta)[y-\eta]-\tau'(\theta)[t-\tau]},$$

where l, m, n satisfy the relation $l^2+m^2=n^2$.

25. Show that wave-functions of type $\gamma f(\theta)$ may be derived from solutions of Laplace's equation of this type by means of the results given on pp. 111, 114. Hence show that there are wave-functions of type $\gamma f(\theta)$ which are not particular cases of a more general wave-function of type $\gamma f(a, \beta)$ where a, β are defined by equations (280).

LIST OF AUTHORS QUOTED

INDEX

(The numbers refer to the pages)

Printed in the United States
By Bookmasters